Cleaner Cars:

The History and Technology of Emission Control Since the 1960s

Other SAE books on this topic

Emissions and Air Quality
by Hans Peter Lenz and Christian Cozzarini
(Order No. R-237)

Automobiles and Pollution
by Paul Degobert
(Order No. R-150)

For more information or to order this book, contact SAE at 400 Commonwealth
Drive, Warrendale, PA 15096-0001; (724)776-4970; fax (724)776-0790;
e-mail: publications @sae.org; web site: www.sae.org/BOOKSTORE.

Cleaner Cars:

The History and Technology of Emission Control Since the 1960s

J. Robert Mondt

Society of Automotive Engineers, Inc.
Warrendale, Pa.

Library of Congress Cataloging-in-Publication Data

Mondt, J. Robert
   Cleaner cars : the history and technology of emission control since the
1960s / J. Robert Mondt.
      p.   cm.
   Includes bibliographical references and index.
   ISBN 0-7680-0222-2
   1. Automobiles—Pollution control devices—History. I. Title.

TL214.P6 M64 2000
629.25'28'0973—dc21   99-057250

Copyright © 2000   Society of Automotive Engineers, Inc.
                            400 Commonwealth Drive
                            Warrendale, PA 15096-0001 U.S.A.
                            Phone: (724)776-4841
                            Fax: (724)776-5760
                            E-mail: publications@sae.org
                            http://www.sae.org

ISBN 0-7680-0222-2

SAE Order No. R-226

Contents

Appendices

Foreword

This book is an historical overview of the evolution of emission controls for automobiles. It is written from my perspective as a participant in the research and development program at General Motors, where I worked for 43 years. As a mechanical engineer, my technical specialties include thermodynamics, heat transfer, and fluid mechanics. My hands-on experience has been in integrating components into complete vehicle systems, with a particular emphasis on introducing the catalytic converter as a primary component for controlling exhaust emissions from automobiles.

I witnessed the evolution of emission controls starting with the environmental concerns in California in the 1950s, and the crusade to eliminate smog. This movement persisted and gained momentum until, finally, the entire auto industry was involved in meeting emission regulations. The conflict and compassion between lawmakers and the auto industry was fascinating to me. Both realized that clean air was important, and both have worked diligently in their specific arenas to achieve improved air quality.

During the past 30 years, somewhat as a hobby, I have accumulated many references, articles, press releases, and stories about the government agencies, scientists, educators, auto industry executives, and auto engineers who appeared as this fascinating drama unfolded. And in my job at General Motors, I was intimately involved in this drama. As more and more organizations in the auto industry, and especially GM, became involved, I was asked to document the historical developments so that we would have available some educational materials for newly hired individuals who were about to become involved in auto emission controls. This was necessary because at the time there was no publication covering the subject overall, particularly none which presented a comparison of vehicle systems as they evolved.

In 1982, I wrote a tutorial called, "An Historical Overview of Emission-Control Techniques for Spark-Ignition Engines."* This proved to be very popular both within GM and outside. Within GM, it was the basis for a training course that was presented more than 20 times. It was also the basis

---

* Mondt, J.R., "An Historical Overview of Emission-Control Techniques for Spark-Ignition Engines," GMR Research Publication 4228, General Montors Research Laboratories, Warren MI, 1982.

for lectures to graduate mechanical engineering students at Stanford and Michigan State for 12 consecutive years, and was published by ASME in 1989 as part of a book entitled "History of the Internal Combustion Engine." Much of this tutorial is also incorporated into the book you now hold in your hands.

The information contained in this book is true and accurate to the best of my knowledge. Any opinions expressed are wholly my own. My perspective is necessarily formed by my role at General Motors; however, to make the viewpoint as general as possible, I have included information provided by other car companies, and I acknowledge the many contributions of individuals outside of General Motors to the development of emission-control technology.

J.R. Mondt

Preface

This book documents the truly amazing story of how we as a nation, over a period of 35 years, succeeded in controlling air pollution from passenger cars. It is dedicated to the hundreds of engineers, scientists, technicians, mechanics, chemists, physical chemists, materials suppliers, university researchers, legislators, and other individuals who contributed to this success story.

The people of the United States should be proud of the U.S. auto industry, manufacturers, and suppliers alike, for the high degree of success they have achieved in controlling air pollution from vehicles. By virtue of a democratic system that permits citizens to voice concerns and motivate government officials to act, industries given the freedom to innovate to find solutions to problems, and an educational system that produces dedicated, hard-working engineers, scientists, technicians, and assembly workers, the air pollution from automobiles as of the year 2000 will have been lowered to minuscule levels compared with those of the late 1960s. Between the years 1965 and 2001, allowable air pollution from passenger cars has been lowered steadily to levels less than 5% of those for pre-control era vehicles; and, at the same time, fleet miles per gallon have more than doubled.

Although the story that follows is one of dramatic progress, now is not the time to rest on our laurels; we must continue to work together to control emissions from automobiles and to ensure a healthy environment for future generations.

Acknowledgments

Many thanks to the following individuals for their contributions to this book and to the field of emission control technology:

Dan Wendland, of GM Research and Development, for editing much of the text, and for most of the technology on converter pressure loss, especially techniques for treating flow maldistribution in the substrate, and for correlating pressure drops for components in the exhaust system. The industry is indebted to Dan for his work in this area.

Kathy Taylor, also of GM Research, who contributed much to catalyst sections.

Harold Haskew, of GM Proving Grounds, who reviewed sections of the text related to vehicle testing and certification, including testing of vehicles in customers use.

Steve Mahan, of GM Delphi, Engine and Engine Management Systems, who reviewed the text.

Bob McCabe, of Ford Motor Co.; Mike Brady, of Chrysler Corp.; Bob Farrauto, of Engelhard Corporation; and Phil Willson, for their review and contributions.

Introduction

Advances in mankind's standard of living over the last century can be characterized in large part by both increased consumption of energy and depletion of energy sources, predominately oil and coal. A byproduct of this process has been a significant increase in the amount of waste matter generated, both on the ground and in the atmosphere. Until recent times, ongoing natural chemical processes have been sufficient to cleanse the environment of most of the pollution and waste materials that have resulted from energy production and consumption. However, in light of a constantly increasing population worldwide and the resultant use of more and more energy-consuming appliances and vehicles, the eradication of these wastes can no longer be left to Mother Nature alone.

At the beginning of the twentieth century, most pollutants still originated from natural sources. However, as early as the 1930s, "smog," originally believed to be a combination of smoke and fog, was present in the Southern California Basin. In other forested regions of the United States, contaminated air resulted in mountain ranges being given such names as the "Blue" or "Smoky" mountains; the chemical forces that produced the colored haze resulted from the reactions of volatile organic compounds (VOCs)—mostly hydrocarbons—and biogenic oxides of nitrogen (NO_X).

By the 1950s, man's contributions to environmental pollution began to increase at an alarming rate, especially in energy-consuming nations such as the United States. At mid-century, major sources of air pollution included human and animal waste, effluents from manufacturing plants and power generating plants, and, increasingly, exhaust from automobiles.

During the prosperous 1960s, the automobile became synonymous with the free-wheeling nature of life and culture in the United States. A prosperous household wanted and could afford to have at least two cars. And the vast freeway system initiated during the Eisenhower Administration made transportation by car convenient and inexpensive, enabling motorists to visit every corner of the United States as well as venture into Canada and Mexico. The U.S. automobile industry could do no wrong! More and more cars were manufactured to satisfy every customer need. New styles of vehicles were produced, including larger models of station wagons, sports cars, and small trucks. Customers responded readily to each advance in comfort and convenience, which included air conditioning, power brakes, power steering, electric windows, and electric door locks. However, the downside was that all of these developments increased the per-vehicle energy consumption.

In the decade of the 1960s, approximately 20% of the energy consumed in the United States could be attributed to automobiles. Essentially all passenger vehicles were powered by hydrocarbon-based fuel, mostly gasoline. As the total automotive fleet expanded rapidly in the 1940s and 1950s, vehicles came under severe attack as air "polluters," especially in California. Exhaust emissions from automobiles increasingly were identified as a significant contributor to air pollution. Professor Phillip Myers of the University of Wisconsin stated, in 1966, that "air pollution is a problem to be concerned with—we automotive engineers have a special interest and contributions to make since automobiles produce about 60% of the total mass of air pollutants" [I1.1].

The engineering and technical communities were challenged to become "good neighbors" of the environment by finding ways to reduce automotive emissions and thereby help to improve air quality. As is frequently the case with social and technological change, the initiative for action started in America's most populous western state, California.

Reference

I1.1   Myers, P.S. "Automobile Emissions—A Study in Environmental Benefits Versus Technological Costs," SAE Paper No. 700182, Society of Automotive Engineers, Warrendale, Pa., 1970.

"ODE"(r) of Pollution

The following poem was written by Ernest W. Landen from Caterpillar Tractor Co. and was printed in the April 1968 issue of the *SAE Journal*. The poem captures the essence of the historical efforts to control pollutants and emissions:

In recent years men of scientific mind
Have been diligently trying to find
A simple and unequivocal solution
To a problem known as air pollution

In years gone by when streets were mud
Enriched with animal droppings and other crud
Debate ensued, mixed with indecent talk
Until villages installed the first sidewalk

But still when people crossed the street
The tenacious muck stuck to their feet,
The fair sex naturally with distaste of grime
Demanded ways to escape this slime.

Some townsmen said with great remorse
We must get rid of the nasty horse,
While others thought dignity could be saved
By simply having the whole street paved!

Thus emissions were limited to discrete piles
Around which people walked with smiles
And busy tumble bugs enjoyed their day
Until the cleaner swept it away.

The people now had settled down
To enjoy the pure fresh air of town
Until the experiments of Professor Otto
Produces a vehicle called the auto.

This faster mode of transportation
Quickly reduced the equestrian population
Where once a single span of horses stood
A hundred now hid beneath the hood.

Where clip clop of horses once filled the street
The roar of tailpipes now set the beat
Autos, taxis, trucks and the bus
All clutter the street in front of us
The products of combustion now of course
Differ completely from those of the horse

Chapter 1

California Crusade to Control Emissions

Even before the Second World War, a cloud was looming on the horizon—the California horizon, to be specific—which ultimately would drastically change the course of the automobile industry, not only in the United States, but worldwide. That cloud was identified and named "smog." People in Southern California gradually became aware that the hazy brownish-white cloud they lived with was becoming more dense, increasingly causing burning eyes and respiratory problems.

Atmospheric conditions in the Los Angeles Basin combined with the nature of the terrain made the region especially suited to the accumulation of air pollutants and the chemical reactions that later were identified as the ones that produce smog. In the Southern California Basin, atmospheric air currents flow easterly from the Pacific Ocean, and when they encounter the mountain ranges east of the basin, this flow is "dammed." This creates a stagnant atmospheric mass, which is ripe for inversion when warm air from the populated urban center rises and traps the pollutants, providing an environment that is conducive to the formation of smog. The chemical reactions responsible for smog formation are described in more detail in Chapter 2.

In 1946, the South Coast Air Quality Management District was formed in Southern California, followed in 1947 by the Los Angeles County Air Pollution Control District. Regulations aimed at controlling air pollution were drawn up and enforced. These required modifications to the petroleum industries, and prohibited outside trash burning and open burning in agriculture. Between 1948 and 1955, the petroleum industry spent $50,000,000 on pollution controls; however, the improvement in air quality was only marginal. As a consequence, in 1955, California set up the Bureau of Air Sanitation.

The Bureau of Air Sanitation set the state's first air quality standards and identified the pollution levels at which the health of some of its most susceptible citizens might be endangered. In 1960, the Motor Vehicle Control Board was formed to regulate automobile tailpipe emissions. This board was responsible for implementing controls of automotive emissions to meet standards for exhaust and crankcase emissions established by the State

---

The Auto Industry Meets California

What was most likely the first official contact between auto industry representatives and the citizenry of California took place in January 1954. A visit to Southern California, documented by C. Heinen and W. Fagley [S1.1], was organized to acquaint the auto industry with the undesirable "brownish cloud" hovering over Southern California. A group of auto engineers and scientists representing the four members of the U.S. auto industry, General Motors, Ford, Chrysler, and American Motors, assembled in Detroit for a train trip to California. John Campbell, Technical Director from the GM Research Laboratories, was chairman of the group, which was a subcommittee of the Automobile Manufacturers Association (AMA). Campbell was well respected in the industry, with a reputation in the field of combustion in internal combustion engines. As described by Charles Heinen, "John had an idealistic attitude about the job—and inspired the group with a sense of mission and eagerness."

During the two-day train trip the contingent reviewed the meager data that was available on the subject of air pollution, exchanged information, and expressed optimism that a solution or a proposal for a solution to the California environmental problem could be agreed upon. The group was aware of theories that the brown cloud was the result of photochemical reactions of gases, but believed that the exhaust from automobiles contributed little to these reactions. The "brown cloud" was thought to be a combination of smoke and fog, thus the term "smog."

The day the auto industry representatives arrived, an impromptu press conference revealed that Californians "viewed the industry both as the villains and the resultant saviors." The auto engineers, in turn, blamed

---

Department of Health [1.2]. In 1964, the California Board approved several emission-control devices for installation on vehicles by supplier companies. These devices included a "direct-flame afterburner" and several catalytic converter designs. All of the devices, however, proved inadequate when tested on vehicles; they either failed to control emissions or caused vehicle driveability problems.

incinerators as the cause of most of the smog. Thus, the stage was set for a confrontational relationship between California and the auto industry.

The AMA subcommittee spent three weeks in California for the purpose of experiencing smog, attending press conferences, and hearing presentations on the subject of air pollution. All individuals involved were sincere in their desire to help find a solution to the problem. California business leaders in attendance included chairmen of boards, and company presidents and vice-presidents, joined by elected state officials. The participation of so many major business interests in the scheduled activities underscored the importance that all parties placed on solving the smog problem in California.

When the contingent returned from California, a flurry of activity was initiated by the auto companies. After intensive reviews with industry experts, the only overall conclusion reached was that the engineering knowledge was not available to provide a device to solve the California smog problem! The auto companies then agreed to pursue a joint program of cooperative research and development to expedite the search for technologies to control air pollution from automobiles. This program resulted in new technologies to reduce emissions from motor vehicles, for example, crankcase ventilation systems and air injection systems. The program continued for fifteen years until it was terminated by antitrust litigation in 1969.

Reference

S1.1   Heinen, C. M. and Fagley, W. S., "Smog—The Learning Years—Building the 88th Story," SAE Paper No. 890813, 1989.

In 1968, two years before the U.S. Congress established the Environmental Protection Agency (EPA), the two agencies in California were combined to form the Air Resources Board. The name was later changed to the California Air Resources Board (CARB). The California legislature saddled CARB with the responsibility of adopting the most effective statewide emission controls feasible for motor vehicle fuels, a wide range of mobile sources, and consumer products [1.3]. Ever since, CARB has been the watchdog of the automobile industry and an unrelenting advocate of lowered emissions from automobiles. CARB's crusade ultimately impacted emission control efforts in the rest of the United States, and has or will be a factor in all countries of the world as their economies develop.

A study undertaken in California in 1965 indicated that the automobile could be responsible for 80% of the unburned hydrocarbons and 65% of the oxides of nitrogen in the atmosphere [1.4], and that these gases were somehow responsible for the undefined air pollution known as smog. As early as 1952, Dr. Arie Haagen-Smit, a Dutch-born scientist at the California Institute of Technology, was credited with linking gaseous emissions from automobiles to the dingy skies and to smog [1.5]. Continuing chemical studies revealed that a relationship existed between unburned hydrocarbons, oxides of nitrogen, and sunlight. It was found that all of these components working together produced ozone, an aggressive oxidizing agent that is harmful to the human respiratory system. The complex chemical and photochemical processes that produce smog have been studied by many investigators. In the early 1960s, a "smog chamber" (a chamber to produce smog) was designed and built at the GM Research Laboratories in Michigan.

This link between smog and the automobile initiated an unparalleled confrontation between the auto industry and the advocates of clean air. Ever since, California has spearheaded the nationwide effort to establish regulations aimed at regaining the air quality levels existent in the 1940s in Southern California. Other areas of the country with similar concerns about air pollution included the major cities of Dallas, Texas; Denver, Colorado; New York; and Washington, D.C. Influenced by the geographical spread of serious air pollution problems, the U.S. Federal Government finally chose to involve itself in the movement.

An agency was formed in 1970 within the U.S. Department of Health, Education and Welfare (HEW), to establish National Ambient Air Quality Standards (NAAQS) for acceptable levels of major air pollutants. Subsequently, the Environmental Protection Agency (EPA) was created by consolidating the activities and responsibilities of 15 separate federal agencies that already existed under HEW. EPA's role, which took effect on December 1, 1970, was to enforce regulations for air quality.

President Richard Nixon appointed William Ruckelshaus as the first administrator of EPA. Ruckelshaus held this post until April 1973, and his administration covered the turbulent formative years of the new agency. His challenge was to enforce the Clean Air Act, which had been passed by the federal government in 1963. The original Act primarily targeted industrial sources of air pollution; however, in 1965, the Clean Air Act was amended to include automobile emissions as well. For several years, Ruckelshaus spent at least 60% of his time attempting to balance the interests of the federal government with those of the auto industry, which contributes approximately 16% to the gross domestic product of the U.S., in devising regulations that would ensure the reduction of HC, CO, and NO_X emissions from automotive vehicles. The remaining 40% of his time was directed to lowering emissions from other sources, primarily from industries.

The automobile industry faced probably its biggest challenge to date as a regulated industry as it entered the 1970s. Ford Motor Company and Mobil Oil Corporation responded to this challenge by forming the Inter-Industry Emission Control Program (IIEC) in April 1967 as a cooperative effort between the auto and oil industries [1.6]. The objective of IIEC was to develop a powerplant and emission-control system that not only lowered emissions, but also improved fuel economy, vehicle driveability, and vehicle durability. As it evolved, IIEC split into two phases, the first beginning in 1967 and the second in 1974. When the second phase was initiated, the acronyms IIEC1 and IIEC2 were instituted to identify the two distinct phases. The overall IIEC program, including both phases, had total expenditures of approximately $32 million and remained in effect until 1977, when it was disbanded by mutual agreement of all parties concerned. By that time, individual industrial companies had developed their own in-house capabilities and felt that further cooperative efforts would provide only diminishing returns. (IIEC1 and IIEC2 are discussed further in Chapter 4-6).

From these beginnings, as a result of the concentrated effort of thousands of individuals, including regulators and industry representatives, the level of automotive exhaust pollutants in the United States has been lowered to seemingly impossible levels. Beginning with the1999 model year and due for completion in 2001, National Low Emissions Levels (NLEV) are being implemented in 46 of the 50 states. The aim is to lower unburned hydrocarbons by 99.3%; carbon monoxide by 96.0%; and oxides of nitrogen by 95.1% from the benchmarks of the uncontrolled levels of the late 1960s. In the remaining four states, one of which is California, unburned hydrocarbons and carbon monoxide are to be lowered to a greater degree, by 99.6% and 98.0%, respectively. These figures are testament to the fact that air quality in the United States, especially in large urban areas, improved markedly in the latter part of the twentieth century. (Progress in improving air quality will be discussed in more detail in Chapter 9.)

References

1.1   Myers, P.S. "Automobile Emissions—A Study in Environmental Benefits Versus Technological Costs," SAE Paper No. 700182, Society of Automotive Engineers, Warrendale, Pa., 1970.

1.2   Jensen, D.A. "The Public's Role in the Automobile Exhaust Emissions Program," SAE Paper No. 660103, Society of Automotive Engineers, Warrendale, Pa., 1966.

1.3   CARB, "California Resources Board Press Release," 1968.

1.4   MacGregor, J.R., "Rational Attack on Smog, Today's Major Automotive Technical Challenge," *SAE Journal*, Vol. 74, No. 1, p. 84, January 1966.

1.5   Haagen-Smit, A.J., "Chemistry and Physiology of Los Angeles Smog," *Ind. Eng. Chem.*, Vol. 44, p. 1342, 1954.

1.6   McCabe, L.J. and Koel, W.J., "The Inter-Industry Emission Control Program—Eleven Years of Progress in Automotive Emissions and Fuel Economy Research," SAE SP-431, *Inter-Industry Emission Control Program 2 (IIEC-2) Progress Report No. 2*, Society of Automotive Engineers, Warrendale Pa., 1978.

Chapter 2

The Air We Breathe

The atmosphere surrounding the earth is a mixture of gases, composed primarily of 21% oxygen and 78% nitrogen. The remaining 1% or so is composed of many other gases, including water vapor, argon, carbon dioxide, hydrogen, neon, helium, krypton and xenon. The total mass of the atmosphere is approximately 5.7×10^{15} tons. Overall, proportions of the various components of air vary little from one place to another. Local imbalances in constituent quantities, which may exist at a specific time, are soon destroyed by huge weather systems, especially turbulent winds that mix the air continuously. The nature of this turbulence may seem surprising, but it is a result of many factors acting together to maintain the balance of the mixture.

The mixing of air between the Northern and Southern Hemispheres is a very slow process. In the Northern Hemisphere, man's pursuit of an improved standard of living has consumed energy at a high rate, creating increased levels of air pollutants. Thus, air pollution in the Northern Hemisphere is more serious than in the Southern Hemisphere. The concentration of air pollutants, especially in heavily populated areas, is aggravated by stagnant air masses, such as in the Southern California Basin, which do not allow air currents to disperse the locally contributed pollutants.

The total amount of carbon dioxide in the atmosphere surrounding the earth has intrigued scientists for years because there are so many processes that contribute carbon dioxide to the atmosphere. The main ones are the respiration of humans and animals, the decay of vegetation, and combustion. At the same time, plants absorb carbon dioxide from the atmosphere, and from it synthesize starch, sugar, and cellulose. In addition, many millions of tons of carbon dioxide are removed from the air by chemical changes initiated when carbon dioxide dissolves in water, either directly or during rainstorms. Aquatic organisms in bodies of water convert dissolved carbon dioxide into calcium carbonate, which is used in these organisms to form shells and skeletons.

Complete combustion of hydrocarbon fuel in an automobile engine adds both water vapor and carbon dioxide to the atmosphere. Incomplete combustion adds such gases as carbon monoxide, nitrogen oxide, nitrogen dioxide, ozone, sulfur dioxide, hydrogen, and hydrogen sulfide; as well as innumerable classes of unburned hydrocarbons. In addition to gaseous emissions, vehicles powered by hydrocarbon fuels emit particulates in the exhaust.

What is Smog?

The are two principal types of atmospheric pollution that have received much attention: "London fog" and "Los Angeles smog." Neither of these should be confused with fog, which consists of extremely small particles of condensed water vapor suspended in air.

"London fog," or London-type air pollution results from smoke and fumes originating from the combustion of hydrocarbon-based fuel. It is produced by industrial processes, power plants, refuse disposal, and household heating systems. The primary ingredients are suspended carbon particulates and sulfur dioxide. Lack of air circulation in the atmosphere permits the buildup of these pollutants, particularly when the weather is still, cool, and damp, such as it frequently is at night. London fog was particularly oppressive during the coal-burning era in London. Conditions similar to "London fog" could be found in other European industrial cities during years when coal was the primary all-purpose fuel. In modern times, this condition has been significantly reduced as a result of the substitution of natural gas and oil as sources of energy combined with the implementation of exhaust emission controls by industries. The automobile contributes very little to this particular type of air pollution.

Los Angeles-type smog air pollution, characterized by the "brown cloud," a reduction in visibility, and a unique odor, results from many chemical and photochemical reactions [2.1, 2.2, and 2.3]. Los Angeles-type smog is not unique to California. Many cities in the United States, including New York, Philadelphia, Chicago, Washington, Philadelphia, Baltimore, Pittsburgh, Cincinnati, Houston, Milwaukee, Atlanta, Dallas-Fort Worth, and Denver have smog air pollution, as do many foreign cities, such as Mexico City, Sao Paulo, Rome, Tokyo, and Frankfurt.

One of the primary concerns about smog among health care providers world-wide is its ozone content. Ozone (O_3) is an aggressive oxidant that causes eye irritation, coughing, chest discomfort, upper respiratory discomfort, and reduced pulmonary function [2.4]. Ozone is formed from the combination of oxides of nitrogen (NO_X), with unburned hydrocarbons (HC), or volatile organic compounds (VOCs), in the presence of sunlight. Sunlight provides the irradiation that excites or "energizes" the oxygen to form ozone. The reactions necessary to produce smog are complex. As shown in Fig. 2-1, they involve peroxy and hydroxyl radicals.

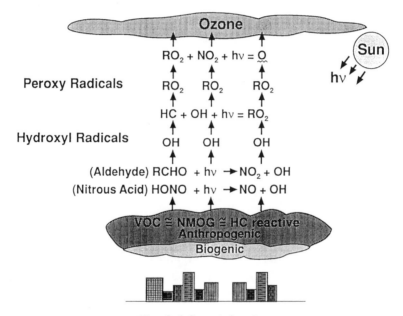

Fig. 2-1 Smog chemistry.

It has taken many years and many man-hours of effort by dedicated scientists to unravel the complex chemical reactions, intermediate reactions, and inter-actions with reactive radicals that ultimately produce ozone. A simplistic summary is that air containing hydrocarbons and oxides of nitrogen reacts under the influence of sunlight, ultimately producing ozone. These complex reactions depend on the amount of hydrocarbons present, the amount of oxides of nitrogen present, and the strength of the sunlight. The reader is referred to References 2.2 and 2.3 for a complete listing of these reactions.

The reactions that produce ozone occur slowly. The mass of smog reactants must remain in place for several hours in order for these reactions to reach equilibrium. Lack of natural ventilation, such as occurs in the Los Angeles Basin, therefore aggravates the smog condition. In the California basin, the prevailing movement of air is from west to east, and during a weather period when there is very little air motion, the air mass all too frequently stagnates, or become "trapped," over the basin.

Hydrocarbons, Carbon Monoxide, and Oxides of Nitrogen

The "conventional" automobile powered by a gasoline engine produces a mix of hydrocarbons, carbon monoxide, and oxides of nitrogen, and hence is a major contributor to smog. Uncontrolled burning of wood in household fireplaces has also been identified as a significant contributor to air pollution, based on a study of Denver's own "brown cloud." In this study [2.5], reported by the General Motors Research Laboratories, ambient air constituents were monitored at seven geographic sites in 1960, using GM's Mobil Atmospheric Laboratory. The results of the study identified the relative contributions of different sources to the Denver "brown cloud": 34% from coal and oil industrial combustion; 27% from motor vehicles; 18% from wood-burning; 12% from natural gas combustion; and the remaining 10% from miscellaneous sources.

Exhaust from an automobile engine burning a hydrocarbon fuel contains at least 150 hydrocarbon (HC) or volatile organic compounds (VOCs). These hydrocarbon compounds have different reactivities, that is, different propensities to react to form ozone. For example in a typical mixture of hydrocarbons in exhaust from an automobile powered by gasoline, one compound, ethylene, accounts for 17% of the reactivity. The olefin family of hydrocarbons, consisting of ethylene, propylene, 1-butene, 1-pentene, 1,3-butadiene, and isoprene, is responsible for approximately 75% of total reactivity. A list of several hydrocarbon species and their reactivities is presented in Table 2-1. The reactivity adjustment factor (RAF) is an arbitrary scale, defined by Dr. William Carter, which rates the relative ozone-producing potential of a specific hydrocarbon.

Table 2-1. Sample of Monitored Hydrocarbons

Species	Reactivity Adjustment Factor	Origin
Benzene	0.28	Unburned fuel
Octane	0.93	Unburned fuel
TM-Benzene	8.83	Unburned fuel
Xylene	8.15	Unburned and partial burn
Toluene	2.73	Unburned and partial burn
Ethene	7.29	Partial burned fuel
Propene	9.40	Partial burned fuel
2M-Propene	5.31	Partial burned fuel
Butadiene	7.7	Partial burned fuel
Methane	0.0148	Partial burned fuel

Volatile organic compounds (VOCs) produced naturally by plant and animal life are identified as biogenic; those produced by civilization are identified as anthropogenic. Motor vehicles, petroleum refineries, chemical plants, gasoline stations, dry cleaners, print shops, aircraft, and all gasoline-fueled off-road and utility engines emit anthropogenic VOCs.

In urban areas, as much as 90% of the carbon monoxide (CO) in local ambient air may be from motor vehicle emissions. Carbon monoxide is produced from the incomplete oxidation of carbon during combustion of hydrocarbon fuels. (With adequate oxygen and a sufficient residence time for complete combustion, all carbon would be oxidized to carbon dioxide.) CO is a highly poisonous gas that hinders oxygen transport from blood tissues, requiring the heart to pump more blood to deliver the same amount of oxygen. Exposure to low levels of CO poses a very real problem for individuals with weak hearts.

Oxides of nitrogen, primarily NO and NO_2, and identified as NO_X, are byproducts of fossil fuel combustion; industrial processing, oil refineries, and power plants are all significant producers of NO_X. Formation of NO_X results from high-temperature (>1500°C) thermal reaction between N_2 and O_2 in air. However, a recent study by the EPA reports that the automobile contributes as much as 50% of the man-made NO_X in the atmosphere [2.6].

NO_X gas has adverse effects on human health, causing increased susceptibility to respiratory infection and decreased pulmonary function. However, the current industry focus on lowering NO_X is because it is a critical constituent in the formation of smog [2.7].

Ozone concentrations are usually small adjacent to roadways. The gaseous reactants that form smog have the same density as air, so they are carried along by atmospheric air mass movements. As the air mass moves, chemical reactions within it continue to take place for a duration of several hours. Consequently, the impact of smog and ozone air pollution may be felt many miles from the source of the pollutants, as illustrated in Fig. 2-2. For this reason, it is essential that state governments, the federal government, and pollution-control agencies cooperate to control this type of air pollution.

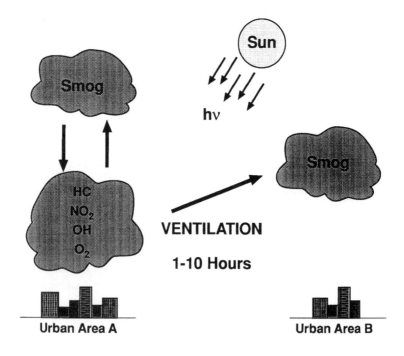

Fig. 2-2 Air movement.

Evidence of the time-delayed and long-distance impact of smog reactions can be seen in the alarming ozone levels in the northeastern region of the United States. Air pollution in this area can be attributed to gaseous emissions produced in the central regions of the United States and carried by the natural movement of air from west to east. Air moving into such a region can become further polluted by local sources; moreover, as noted, some increase in pollution occurs as the ozone produced in the moving air continues to undergo chemical reactions with ongoing exposure to sunlight.

In response to studies detailing these findings, in 1990 the U.S. Congress, as part of the Revised Clean Air Act of 1990, formed the Ozone Transport Commission (OTC). The OTC includes twelve northeastern states and the District of Columbia. The objective of this consortium is to improve air quality in the northeastern United States.

The Ozone Transport Commission in the Northeast petitioned the EPA for the authority to implement the stringent California emission standards. This petition was required because the Revised Clean Air Act specifically allowed only California to have a set of emission standards that differed from federal standards. EPA approved this petition in 1994.

1989 Study to Understand Ozone Chemistry

In March 1989, members of California's South Coast Air Quality Management District voted in a sweeping plan to control smog and its most serious component, ozone. Sponsors were the California Air Resources Board (CARB), Coordinating Research Council, and the Motor Vehicle Manufacturers Association. This action was prompted by air quality measurements in Los Angeles showing unusually high concentrations of ozone. For several years, on one-third of the days during July, ozone concentrations exceeded two times the standard of 120 ppb. To further identify and analyze the nature of the chemistry involved in producing ozone, industry research organizations and government agencies established the Southern California Air Quality Study.

Experiments in Ozone Formation

In 1987, GM Researchers set out to determine the role of hydrocarbons (HC) and oxides of nitrogen (NO_X) in ozone formation. Two sampling sites were set up in California, one in downtown Los Angeles, near Dodger Stadium, and one 30 miles east, in Claremont. At each site eight portable 500-liter volume Teflon bag smog chambers were set up. In the morning these transparent chambers were filled with ambient "air," and then exposed to sunlight to "cook" during the day. Different proportions of HC, ambient air, and NO_X were mixed to achieve levels of HC and NO_X that were 25 to 50% of the levels existing in the ambient air mixture. Ozone formation in these mixtures was then measured and the results were correlated with concentrations of HC and NO_X. The experiments were repeated, and the interactions between HC and NO_X were statistically analyzed. Based on these results, an empirical mathematical model

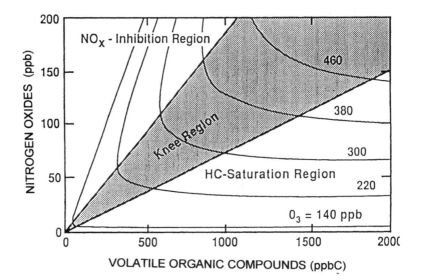

Fig. S2-1 VOC, NO_X, and O_3 interactions. (Source: [Ref. S2.2].)

was developed which correlates the relative influence of the various factors contributing to ozone production. Fig. S2-1 summarizes the results of matched experiments performed at the two sites.

This graph shows constant ozone concentrations (isopleths) as a function of NO_X and HC concentrations [S2.1, S2.2]. For large concentrations of HC, lowering NO_X concentration substantially lowers ozone concentration. For high concentrations of NO_X, lowering HC concentration substantially lowers ozone concentration. In the knee region, which is typical for Los Angeles air, lowering both NO_X and HC concentrations is the best solution to lowering ozone concentration.

This graph indicates that care must be exercised in prescribing the best approach to lowering ozone levels because this will depend on the constituents present in the ambient air. The benefits of reducing the atmospheric concentrations of nitrogen-bearing species through reductions in NO_X emissions, and of reducing secondary organic aerosols through reductions in HC or VOC (volatile organic compound) emissions must be factored into the analysis. Another important point is that the positions of the ozone isopleths in this graph may not be the same as they would be in other urban areas, so the use of a single plot to define control strategies for all locations may be an oversimplification.

References

S2.1   Seinfeld, J.H., "Urban Air Pollution: State of the Science," *Science*, Vol. 243, February, 1989.

S2.2   Kelly, N.A. and Gunst, R.F., "Response of Ozone to Changes in Hydrocarbon and Nitrogen Oxide Concentrations in Outdoor Smog Chambers Filled with Los Angeles Air," *Atmospheric Environment*, Vol. 24a, No. 12, pp. 2991–3005, 1990.

National Ambient Air Quality Standards (NAAQS) 1970

When the Clean Air Act was amended in 1970 [2.9], national standards for air quality were promulgated by the U.S. Congress. Designated the National Ambient Air Quality Standards (NAAQS), these were established to protect the health and welfare of the American citizenry. Subsequently, during a long and tedious process involving the review of hundreds of documents on the health effects of pollutants in ambient air, a National Air Quality Advisory Committee was formed to work with the Environmental Protection Agency (EPA). The Clean Air Act sanctioned the EPA as the responsible agency to implement the NAAQS standards, but designated the states as the enforcing bodies. EPA either grants funding to, or withholds funding from, selected states with nonattaining air pollution regions. EPA also supports research studies to develop additional scientific data to update these standards.

The NAAQS standards are not based wholly on scientific measures; some elements of personal judgement were included in setting them. However, they represent a reasonable and prudent definition of the air pollution control necessary to ensure adequate protection of public health in the United States [2.10].

Air pollutants from automobiles consist primarily of particulates, carbon monoxide, hydrocarbons, oxides of nitrogen, and ozone, which is produced by the reactions described above. Limits for these pollutants are as follows [2.11]:

- Ozone, O_3, target: 120 ppb (target 1975). Non-attainment if 120 ppb exceeded one day each year for a period of three years.
- CO: 9 ppm for 8 hours, 35 ppm for 1 hour.
- NO_2: 5 ppm.
- H_2S, threshold limit: 10 ppm; short term: 200 ppm for 10 min, 100 ppm for 30 min, and 50 ppm for 60 min.

(The values listed above were confirmed with passage of the Revised Clean Air Act of 1990, discussed in Chapter 7.)

In July 1997, EPA adopted a revision to the NAAQS ambient air quality standard for particulate matter. This revision includes particulate matter as small as 2.5 microns, and set the annual value at 15 micrograms per cubic meter, and the 24-hour standard at 65 micrograms per cubic meter. These standards replace previous values for particulate matter as small as 10 microns, which were an annual average of 50 micrograms per cubic meter and a 24-hour standard of 150 micrograms per cubic meter.

Clean Air Act Classifications for Urban Areas

Based on the average air quality measured by the individual states between 1987 and 1989, the Revised Clean Air Act of 1990 established five categories for substandard air quality in urban areas: *marginal, moderate, serious, severe,* and *extreme.* Urban areas in which measured levels of air pollutants exceeded the health standards were designated as nonattainment regions, and were listed in the U.S. Federal Register. Cities classified as marginal, such as Reno, were to meet the standard in 1995. Areas classified as moderate, such as Dallas-Fort Worth, had a target of 1996 to meet the federal ozone standard (see Fig. 2-3) [2.12].

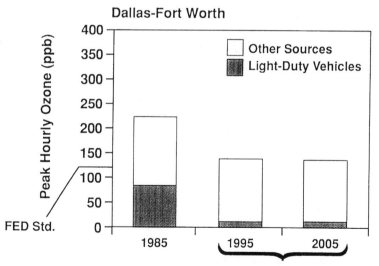

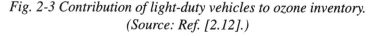

Fig. 2-3 Contribution of light-duty vehicles to ozone inventory. (Source: Ref. [2.12].)

17

Cities classified as serious, such as Washington D.C., were given until 1999. Cities classified as severe, such as New York, have a target of 2007 to meet the federal ozone standard. The Los Angeles Basin, classified as extreme, has a deadline of 2010 to meet the California ozone target (Fig. 2-4). The California target was established prior to passage of the Revised Clean Air Act of 1990, by California's South Coast Air Quality Management District. The objective is to reduce ozone by 50% and lower both HC and NO_X by 80% by 2010.

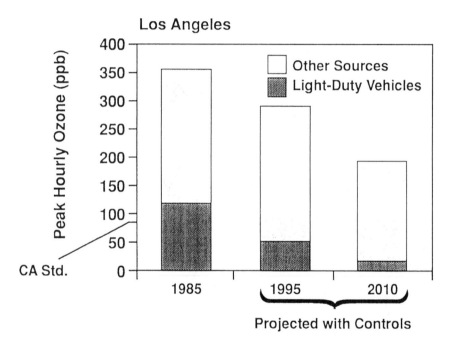

Fig. 2-4 Contribution of light-duty vehicles to ozone inventory.
(Source: Ref. [2.12].)

The road to improved air quality has not been a smooth one. In 1987, sixty urban areas exceeded the tolerable standard for ozone, and by 1993, ninety-eight urban areas exceeded the standard. This nonattainment is due to insufficient reductions in VOCs and oxides of nitrogen. VOC gases in the air have only been reduced by 10% to 25% during the past 25 years instead of the targets of 25% to 50%; and NO_X emissions have not been lowered by the desired 50%.

If cities are to reach the targeted federal ozone levels, emissions from all sources must be lowered, in particular VOCs and NO_X. In this effort, automotive emissions remain a prime concern. A report published in 1991 by the U.S. EPA stated that, in the United States, motor vehicles generate 50% of the CO, 27% of the VOCs, 29% of the NO_X, and 17% of the particulate matter in the atmosphere [2.13]. In addition, John Seinfeld of the California Institute of Technology in 1985 [2.2] reported his findings on the sources of emissions in California. Sources of HC were identified as 40% from vehicles, 30% from solvents, with the remainder from small, industrial, and off-road engines. Sources of NO_X were identified as 45% from vehicles, 30% from utilities, with the remainder from fuel consumption by industries.

All nonattainment urban areas have targets for the reduction of gases in the atmosphere, which fall under the directives of their State Implementation Plans (SIP). However, the targets for California are the most severe. Under the circumstances, EPA has allowed California the privilege of establishing more stringent limits for HC and NO_X emissions from vehicles.

References

2.1   Haagen-Smit, A.J., "Chemistry and Physiology of Los Angeles Smog," *Ind. Eng. Chem.,* Vol. 44, 1954, p. 1342.

2.2   Seinfeld, J.H., "Urban Air Pollution: State of the Science," *Science,* Vol. 243, February, 1989.

2.3   Committee on Tropospheric Ozone Formation and Measurement, *Rethinking the Ozone Problem in Urban and Regional Air Pollution,* National Academy Press, Washington, D.C., 1991.

2.4   National Academy of Sciences, "Ozone and Other Photochemical Oxidants," Committee on Medical and Biological Effects of Environmental Pollutants, 1977.

2.5   Wolff, G.T. et al., "Clearing the Air on Denver's Brown Cloud," *Search,* Vol. 15, No. 6, General Motors Research Laboratories, 1980.

2.6   U.S. EPA, "National Air Quality and Emission Trends Report 1992," EPA, Research Triangle Park, N.C., 1993.

2.7 Moore, C. and Walsh, M.P., "Motor Vehicles and Long Range Transport of Pollutants," SAE Paper No. 851209, Society of Automotive Engineers, Warrendale, Pa., 1985.

2.8 Kelly, N.A. and Gunst, R.F., "Response of Ozone to Changes in Hydrocarbon and Nitrogen Oxide Concentrations in Outdoor Smog Chambers Filled with Los Angeles Air," *Atmospheric Environment*, Vol. 24a, No. 12, pp. 2991–3005, 1990.

2.9 U.S. Government, "National Primary and Secondary Ambient Air Quality Standards," *The Federal Register* 36: 8186, April 30, 1971.

2.10 Heuss, J.M., Nebel, G.J., and Colucci, J.M., "National Air Quality Standards for Automotive Pollutants—A Critical Review," *APCA Journal*, Vol. 221, No. 9, 1971.

2.11 Weiss, G., *Hazardous Chemicals Data Book*, Noyes Data Corporation, Park Ridge N.J., 1980.

2.12 AQIRP, "Auto/Oil Air Quality Improvement Research Phase I Final Report," AQIRP, May 1993.

2.13 U.S. Environmental Protection Agency, Office of Air Quality, Planning and Standards Technical Support Division, "National Air Pollution Emission Estimates: 1940–1990," 1991.

Chapter 3

Automotive Emissions Identified

In the mid-1960s, three pollutants from automobile exhaust were identified for control: hydrocarbons (HC), carbon monoxide (CO), and oxides of nitrogen (NO_X). Emission limits for these gases became law in the U.S. in 1971. Emission limits for particulates were added in 1986, and emission limits for aldehydes (HCOH) were added in 1993.

Combustion of hydrocarbon fuel in an internal-combustion engine involves hundreds of chemical reactions. Gaseous products from this combustion are primarily CO_2, CO, H_2O, H_2, O_2, N_2, NO, NO_2, and many, many HC compounds. The amount of each product varies with the air to fuel ratio (A/F), as shown in Fig. 3-1 [3.1]. A simplistic combustion "equation" is:

Gasoline + Air $\rightarrow$ Combustion $\rightarrow$

(Products of Combustion) $CO_2 + O_2 + N_2 + H_2$

+ (Pollutants) $CO + NO + NO_2 + C_aH_a + \ldots C_nH_n$

The A/F for minimum brake specific fuel consumption (bsfc), depending on the combustion chamber design, usually occurs between an A/F of 16 and 18. For maximum fuel economy, spark ignition engines should be operated at an A/F lean from stoichiometric and depending on the shape of the combustion space in the engine. At the A/F for best economy, HC emissions are at a minimum, CO is quite low, and NO_X is near its maximum. At a fuel-rich A/F, output of CO is high because of excess fuel in the mixture. As fuel content is lowered from a rich mixture, the leaner combustion produces less and less CO, which theoretically should reach zero at stoichiometric A/F.

However, CO production does not decrease to zero even for a lean A/F. This is partly because of cylinder-to-cylinder maldistribution, and also because there is insufficient time to complete the oxidation of CO to CO_2.

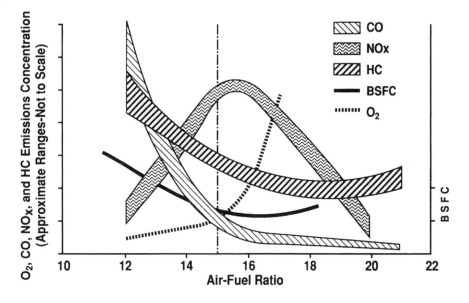

Fig. 3-1 Exhaust products from combustion of hydrocarbon fuels.

The production of unburned HC is also high for fuel-rich burning mixtures because there is insufficient oxygen for the chemical oxidation reaction. HC production decreases as fuel content is lowered or the A/F becomes lean. Like CO production, HC production also does not decrease to zero at stoichiometric A/F.

During a cold start, large quantities of HC and CO are produced from the excess fuel, or fuel enrichment, required to start the engine. Fuel enrichment is necessary during a cold start to provide enough light hydrocarbons to initiate combustion. Large quantities of HC and CO are also produced when the engine is operated under conditions of "power enrichment," for example, for quick accelerations, trailer towing, or hill climbing. Such activities add fuel to the mixture and result in a rich A/F, thereby producing large quantities of HC and CO.

NO_X production from the combustion process peaks at an A/F slightly lean from stoichiometric. This is because production of NO_X is high when combustion temperatures—and thermal efficiencies—are high. NO_X production decreases when combustion temperatures decrease, and this is true for both rich and lean A/F operation. NO_X production also decreases as the A/F is changed in a direction that is either rich or lean from the peak combustion condition; again, this is because the combustion temperature is lowered.

Oxygen levels of 0.5 to 1.0% are usually present in exhaust at a stoichiometric A/F. The amount of oxygen is very low at rich A/F and increases as the A/F becomes more lean. Oxygen obviously must be available for oxidation reactions; and it must be *absent* for reducing reactions!

Since reactive hydrocarbons are key constituents of the chemical processes that produce smog, California has passed legislation limiting HC emissions to levels lower than those required by federal regulations. In addition, beginning in 1994, in California the measurement of total HC emissions was replaced by the measurement of NMOG (non-methane organic gases). This was done to reflect the ozone-forming potential of HC emissions. (Methane is nonreactive in the chemical processes that produce ozone.) Furthermore, separate regulations are in place for five "toxic" hydrocarbons: formaldehyde, acetaldehyde, benzene, 1.3-butadiene, and polycyclic organic matter.

As a result of the new California regulations, smog-forming reactivity must be defined for *each* of at least 150 different HC species that have been identified in the exhaust from a vehicle during the FTP driving schedule (Table 3-1). Table 3-1 lists the "Carter reactivities" for each of the HC species. Carter reactivities, or reactivity factors, provide and arbitrary index of the relative ozone-forming potential of each HC species.

Table 3-1. Hydrocarbons in Exhaust Gas and Carter Reactivities

Name	Reactivity	Name	Reactivity	Name	Reactivity
1,3-Butadiene	7.7	Crotonaldehyde	3.7	Cyclohexane	0.84
1,3,5-TM-Benzene	7.5	N-Butyraldehyde	3.7	Ethanol	0.79
1,2,3-TM-Benzene	7.4	T-4-Octene	3.6	Butanone	0.76
1,2,4-TM-Benzene	7.4	T-2-Octene	3.6	2,3-DM-Butane	0.74
T-2-Butene	7.3	2,4,4-TM-2-Pentene	3.6	Indan	0.73
C-2-Butene	7.3	C-2-Octene	3.6	2,3-DM-Hexane	0.72
1,2,3,4-TetM-Benzene	6.7	3M-Cyclopentane	3.3	2,2,4-TM-Pentane	0.72
1,2,3,5-TetM-Benzene	6.7	Cyclohexene	3.3	3,3-DM-Hexane	0.72
1,2,4,5-TetM-Benzene	6.7	3E-C-2-Pentene	3.2	2,3,4-TM-Pentane	0.72
Propene	6.6	Pentaldehyde	3.2	2,2-DM-Hexane	0.72
2M-1,3-Butadiene	6.5	2M-1-Pentene	3.0	2,4-DM-Hexane	0.72
Formaldehyde	6.2	3M-1-Pentene	3.0	2,5-DM-Hexane	0.72
T-2-Pentane	6.2	4M-1-Pentene	3.0	3M-Heptane	0.72
C-2-Pentane	6.2	1-Hexene	3.0	4M-Heptane	0.72
1-Butene	6.1	3,3-DM-1-Butene	3.0	2M-Heptane	0.70
M&P-Xylene	6.0	Hexanaldehyde	2.67	2,2,5-TM-Hexane	0.68
3M-T-2-Pentene	5.3	Styrene	2.67	2,2-DM-Pentane	0.68
4M-T-2-Pentene	5.3	1-Heptene	2.40	3,5-DM-Heptane	0.68
C-3-Hexene	5.3	3M-1-Hexene	2.40	4M-Octane	0.68
4M-C-2-Pentene	5.3	Toluene	1.90	2,3,5-TM-Hexane	0.68
T-2-Hexene	5.3	2.4.4-TM-1-Pentene	1.90	2,3-DM-Heptane	0.68
Propadine	5.3	E-Benzene	1.80	2,4-DM-Heptane	0.68
Ethene	5.3	M-Cyclopentane	1.70	Pentane	0.64
C-2-Hexene	5.3	Cyclopentane	1.60	Butane	0.64
2M-2-Pentene	5.3	1-Nonene	1.60	Hexane	0.61
1M-3E-Benzene	5.3	Propylbenzene	1.50	2,4-DM-Octane	0.60
1E-2M-Benzene	5.3	T-1,2-DM-Cyclopentane	1.50	2,2-DM-Octane	0.60
T-3-Hexene	5.3	C-1,3-DM-Cyclopentene	1.50	Heptane	0.48
TM-4E-Benzene	5.3	E-Cyclopentane	1.46	MTBE	0.47
0-Xylene	5.2	N-Propylbenzene	1.44	2,2-DM-Butane	0.41
2M-2-Butene	5.0	1,1-DM-Cyclohexane	1.36	Octane	0.41
1,2-DE-Benzene	4.8	E-Cyclohexane	1.36	Methanol	0.40
1,3-DE-Benzene	4.8	C-1,2-DM-Cyclohexane	1.36	Acetone	0.39
1,4-DE-Benzene	4.8	T-1,3-DM-Cyclohexane	1.36	Ethyne	0.37
Acrolein	4.6	T-1,4-DM-Cyclohexane	1.36	2,2,3-TM-Butane	0.35
Propionaldehyde	4.6	C-1,3-DM-Cyclohexane	1.36	Propane	0.33
T-3-Heptene	4.4	ETBE	1.33	Nonane	0.29
T-2-Heptene	4.4	1C,2T-3-TM-Cyclopentane	1.3	Benzene	0.28
C-2-Heptene	4.4	N-Butylbenzene	1.29	Decane	0.25
3M-T-3-Hexene	4.4	S-Butylbenzene	1.29	Propyne	0.24
2,3-DM-2-Pentene	4.4	M-Cyclohexane	1.17	Undecane	0.21
2M-2-Hexene	4.4	2,4-DM-Pentane	1.07	Dodecane	0.19
1M-4-1-Butabenzene	4.3	3,3-DM-Pentane	0.96	2,2-DM-Propane	0.19
1-Pentene	4.2	2,3-DM-Pentane	0.96	2-Butyne	0.18
3M-1-Butene	4.2	3M-Pentane	0.95	1-Butyne	0.18
2M-Propene	4.2	2M-Pentane	0.91	Ethane	0.15
Cyclopentene	4.0	Naphthalene	0.87	Methane	0.01
Cyclopentadiene	4.0	3M-Hexane	0.85	P-Tolualdehyde	-0.47
Acetaldehyde	3.8	2M-Hexane	0.85	Benzaldehyde	-0.54
2M-1-Butene	3.7	2M-Propane	0.85		

Hydrocarbons

Most of the hydrocarbons emitted from a fully warm engine running at a lean A/F have been traced to the cold, thin, stagnant oil film near the walls of the combustion chamber, or to crevices existing between mating parts, such as spark plugs, valve rims, pistons, piston rings, and cylinder walls. Theoretically, complete combustion of the fuel would produce no unburned hydrocarbons. However, in an operating engine, the combustion event is completed in milliseconds—so quickly that the result is incomplete combustion and the production of innumerable species of hydrocarbons.

Besides the effects of incomplete combustion, reactions that oxidize unburned hydrocarbons are cooled to terminate oxidation in the cool boundary layers adjacent to the edges of the combustion space. This cooling quenches the combustion event and results in "end gases" or unburned hydrocarbons. Some of the unburned hydrocarbons are subsequently oxidized by mixing with the bulk gas, some are swept out of the combustion space into the exhaust manifold during blowdown, and some are pumped out during the exhaust stroke. A small amount of hydrocarbons remains trapped in the combustion space and may be oxidized during the next combustion cycle of the engine.

Hydrocarbons discharged into the exhaust manifold may be oxidized before leaving the manifold if the temperature exceeds 750°C, and if the gases are retained in the exhaust manifold in the presence of oxygen for a sufficient residence time [3.2].

As shown in Fig. 3-1, when the fuel-air mixture is leaned excessively, HC levels increase because of incomplete combustion. Ether the mixture is too lean to support combustion, or the temperature is too cool, or both. The "lean limit" identifies the maximum value of A/F ratio for which combustion is stable and complete, resulting in smooth engine operation. For A/F beyond the lean limit, the combustion process degenerates, resulting in an increase in HC output from the engine.

Except for the five toxic hydrocarbons mentioned previously (formaldehyde, acetaldehyde, benzene, 1.3-butadiene, and polycyclic organic matter), small concentrations of hydrocarbon vapors are not toxic unless held in suspension and exposed to sunlight in the presence of oxides of nitrogen, that is, unless conditions are right for the formation of smog. Given the frequency of such conditions, regulated levels for hydrocarbons are based on smog-forming potential.

Prior to the development and implementation of emission controls, unburned hydrocarbons were emitted from crankcase ventilation, carburetor ventilation, fuel tank ventilation, and unburned or partially burned fuel in the exhaust. Crankcase and fuel vapors are essentially 100% unburned hydrocarbons, and accounted for approximately 40% of the unburned hydrocarbons emitted from passenger cars prior to 1970. The unburned or partially burned fuel in the exhaust accounted for the remaining 60% of the unburned hydrocarbons from cars prior to 1970.

Emissions of hydrocarbons from a vented crankcase consist of oil vapors and blowby gases. These gases must be removed from the crankcase to eliminate pressure buildup and the potential danger of explosion of the hydrocarbon-air gas mixture in the crankcase. Blowby gases must also be removed from the crankcase to avoid condensation of combustion products in the lubricating oil, which may produce "sludge." Both condensation and sludge lead to inferior lubrication and excessive wear of engine parts.

An additional source of unburned hydrocarbons is the mixture of hydrocarbon gases and air in a partially filled gasoline fuel tank. Each time the gas mixture is replaced with new fuel while the tank is being filled, the displaced gases are purged from the fuel tank.

Carbon Monoxide

Carbon monoxide (CO) is a toxic gas. All carbon monoxide emitted from an automobile comes from the exhaust pipe. It is formed when inadequate oxygen is present and when combustion does not proceed for long enough to complete the oxidation of all carbon to CO_2. CO is also produced as an intermediate product in the oxidation of hydrocarbons [3.2]. CO discharged from the engine into the exhaust manifold may be oxidized to CO_2 if adequate oxygen is present and the gases remain hot for a sufficient residence time.

Standards for carbon monoxide have been established for both short-term and long term exposure. It is important to avoid the buildup of high concentrations of carbon monoxide in closed spaces. Gasoline service stations, public and private garages, and roadway tunnels must therefore be well ventilated. Without emission controls, crowded downtown regions of large cities could develop pockets of CO when large numbers of vehicles are present.

Oxides of Nitrogen

Oxides of nitrogen are chemical compounds formed by the combination of nitrogen and oxygen under the extremely high temperatures that occur during a combustion event in an internal combustion engine. If left for several hours, these compounds will naturally decompose back into oxygen and nitrogen. However, because of the short duration of the combustion event, <10 milliseconds, followed by quenching of the exhaust gases, NO_X (NO + NO_2) remains in the exhaust. As described in Chapter 2, NO_X plays a key role in the chemical and photochemical reactions that produce smog. It is therefore important to control NO_X output from automotive engines.

Various techniques may be used to inhibit the formation of NO_X within the combustion chamber. NO_X can be lowered by operating the engine at either a very rich or very lean A/F, or by slowing down the combustion event and reducing the temperature. An inert gas can be injected into the combustion space to slow combustion and reduce peak temperatures. In a technique called exhaust gas recirculation (EGR), which has been very successful, exhaust gas is used as the source of inert gas. (EGR is discussed in much detail in Chapter 5.) NO_X can also be lowered by retarding the spark advance setting. This results in the production of less NO_X in the engine because it lowers the peak flame temperature; however, techniques such as this which lower the temperature carry a penalty of decreased efficiency or increased gasoline consumption.

Particulates

Particulate matter is present in the exhaust of vehicles powered by either spark-ignition or compression-ignition (diesel) engines. It consists of unburned carbon and ash particles, agglomerates of organic particles, soluble organics, sulfates, and miscellaneous oil additives and wear materials. The major source of particulate matter is the agglomeration of carbon into minute particles as the combustion event concludes. Particulates are a health hazard for two reasons: (1) They can be injected into the lungs, deposited there, and remain, ultimately impacting the performance of the lungs, and (2) They can react with other substances to produce harmful effects. For example, sulfur oxide particles can combine with water present in exhaust gas to form sulfuric acid vapor. And carcinogens such as polycyclic aromatic hydrocarbons (PAH), can adsorb onto the carbon particles present in exhaust, which can then be ingested and cause cancer.

A Word on Diesel Engines

Since diesel engines operate at very lean A/F, usually greater than 20/1, excess air is available in the combustion space for nearly complete oxidation of CO and HC. The excess air in a diesel engine lowers combustion temperatures, resulting in the production of less NO_X than from a spark-ignition engine. Thus, diesel combustion produces some HC, low CO, and low NO_X. Unfortunately, NO_X levels in diesel exhaust are not low enough to meet exhaust emission standards. At the same time, excess air in diesel exhaust makes it difficult to use catalysts to reduce the NO_X. However, beginning in the 1980s, oxidizing catalysts have been used to oxidize CO, soluble organic matter, and gaseous HC from diesel engines in trucks, buses and passenger cars. Chapter 5 provides additional information about catalysts for diesel-powered vehicles.

Compared to spark-ignition (Otto-cycle) engines, Diesel-cycle engines produce larger quantities of particulates because the fuel charge is injected into a combustion space that is essentially filled with air. At the end of the combustion event, the flame front is cooled before all the fuel is oxidized, and the unburned carbon in the fuel is oxidized by surrounding air, producing particles or particulates.

References

3.1   Ferguson, C.R., and Kirkpatrick, A.T., Internal Combustion Engines, 2nd Edition, John Wiley and Sons, New York, N.Y., 2000.

3.2   Herrin, R.J., "Emissions Performance of Lean Reactors—Effects of Volume, Configuration, and Heat Loss," *SAE Trans.*, Vol. 78, pp. 31–51, 1978 (SAE Paper No. 780008).

Chapter 4

The Role of Industry and the Role of Government

During the 1950s, environmental activists raised the alarm about air pollution among U.S. citizens. The activist movement was strongest in California where the "brown cloud" identified as smog was clearly having a deleterious impact on the quality of life, especially in the Los Angeles Basin. Dr. Arie J. Haagen-Smit at the California Institute of Technology conducted a series of scientific studies in 1954 that identified a link between the automobile and smog [4.1]. Efforts spearheaded by Dr. Haagen-Smit and his colleagues prompted the U.S. federal government in 1955 to sponsor an investigation into the relationship between the automobile and air pollution.

A blue-ribbon panel was subsequently formed in 1967 under the leadership of Robert Morse, the Department of Commerce's Secretary for Science and Technology [4.2]. It was named the "Panel on Electrically Powered Vehicles," since its primary function was to assess alternative automotive powerplants that would not contribute to air pollution, with an obvious focus on electric-powered vehicles. In addition to electric, alternative propulsion systems to be evaluated included gas turbine, diesel, hybrid, and external combustion (steam and Stirling). This panel was subdivided into four main parts: (1) Air Pollution—The Problems and the Risks; (2) Technology and the Control of Automotive Air Pollution; (3) The Role of Industry; and (4) The Role of Government.

The committee assessing the problems and risks of air pollution concluded that prompt and effective action would be required to control further contamination of the atmosphere. The automobile was charged with being largely responsible for the formation of photochemical smog in Los Angeles and several other metropolitan areas; it was targeted as contributing significantly to the amount of carbon monoxide, oxides of nitrogen, and lead compounds in the air.

This committee also envisioned the role of the U.S. government as that of cataloging the masses of pollutants emitted from all sources and identifying the specific pollutants that could be attributed to the automobile. Acting on the scientific data collected, the government would then establish and enforce emissions standards for automobiles, which would be based on a driving schedule representing a motorist's typical use of a vehicle. The driving schedule chosen would require that the engine and drivetrain be subjected to all essential operating conditions. The committee also recommended that the reliability and durability of projected pollution control devices be demonstrated. It was assumed that some form of local inspection system would probably be required to ensure proper operation of emission control systems over the life of the vehicle. The committee was hopeful that quick, inexpensive, and reliable measuring tests could be developed.

The U.S. government was also expected to establish national air quality standards, which all pollutant-generating industries would be required to meet. It was further expected that the government would assess the impact of specific and combined air pollutants on public health; at the time, a lack of scientific information made it impossible to accurately measure the results of such pollution and forecast its health consequences.

The most significant U.S. government role to date to control pollutants from automobiles has been to establish emission control requirements and implement an enforcement time schedule. This was accomplished for the entire United States while still including aggressive California as part of the national program. In 1963, the U.S. Congress passed the Clean Air Act of 1963. Because of ongoing deliberations, methods for controlling emissions from automobiles were not included in the Act until an amendment was passed in 1965. In 1970, the Environmental Protection Agency was formed as the enforcing body, and the journey toward cleaner air began in earnest. In response to these governmental actions, starting in California in 1966, and nationwide in 1968, vehicles in the United States became subject to meeting emission regulations.

Clean Air Act Legislation

In 1961, the California legislature set up the California State Motor Vehicle Pollution Control Board and established a driving schedule [4.3] that included a cold start and was repeated seven times for each vehicle test. The allowable emissions output from a vehicle when operated on this schedule were established by the California State Health Department as 275 ppm of unburned hydrocarbons and 1.5% carbon monoxide. All equipment was required to have a maintenance-free life of 12,000 miles [4.4]. Beginning with model year 1966, all vehicles delivered for sale in California were required to meet these standards [4.5, 4.6].

Emission controls for automobiles were not accepted without a great deal of stress on the California social scene. There was a flurry of interactions between the public, the press, and the legislature. The "public," consisting primarily of vehicle owners, wanted air pollution to go away, but did not want to have regulations imposed on it, and did not want to pay for solving the problem. The public blamed "big business," and expected big business to pay for cleaning up the air, not realizing that the costs would ultimately be passed on to themselves, the consumers. Massive mailings to members of the California legislature provided opinions and suggested solutions; however, most of the solutions were not technically feasible, and therefore served to confuse the legislators. In one week alone in 1964, the Motor Vehicle Control Board received 2000 letters protesting the cost of automobile anti-smog devices. Reflecting this public agitation, between January and June 1965, thirteen full legislative sessions were devoted solely to the topic of air pollution from automobiles. California Senate Bill 317, one of the first dealing with control of air pollution, was amended 11 times before passage.

In 1966, the U.S. Congress reported the results of studies of environmental pollution [4.7] and came to the realization that the extent and complexities of the matter would ultimately require the establishment of a specific governmental agency to investigate and oversee the problem; and that passage of legislation would likely be necessary, especially since the California Air Resource Board had already been formed in 1968, along with subcommittees to advise the state legislature. So, in 1970, before the process of identifying the myriad of individual chemical processes that generate smog could get underway, the U.S. government formed the Environmental Protection Agency (EPA) to monitor air pollution, set standards, and develop test procedures to control air pollution from automobiles.

U. S. Driving Test Procedures

From the outset, a major problem in designing an effective automotive emissions control system has been establishing a representative test procedure. All parties involved recognized that, to be credible, vehicles would ultimately have to be tested and emissions measured during a driving procedure. Under most operating conditions, vehicles are subjected to much transient operation, including a wide variety of driving trips. To monitor exhaust gases during transient operation, a chassis dynamometer would be required, along with sophisticated sampling and measuring instruments.

The Environmental Protection Agency (EPA)

The EPA is responsible for monitoring and controlling all environmental pollution, not only that produced by vehicles, but also that from industrial, commercial, and community sources. The Office of Program Management has overall responsibility for program development and provides administration, data processing, and laboratory support to all divisions of the EPA. Several divisions interact with vehicle manufacturers. The Emission Control Technology Division is responsible for identifying the nature of emissions from existing mobile sources, developing test procedures, and making technology assessments. The Certification and Surveillance Division reviews and evaluates applications for certification of motor vehicles and engines and develops and conducts surveillance activities with respect to emissions. The Alternative Advanced Power Systems Development Division is responsible for investigating alternative power sources for vehicles.

In 1966, the U.S. federal government (HEW), with minor changes, approved legislation adopting the driving schedule proposed by California, known as the 7-mode cycle [4.8]. In effect, vehicles in the 49 remaining United States had to use the same driving test as California, beginning with the 1968 model year, in response to the Muskie Bill. The major drawback of the 7-mode cycle test procedure was that pollutants in the exhaust stream were measured in terms of concentration, or parts-per-million fractions, rather than mass emissions.

In 1972, an EPA ruling replaced the original "7-mode" cycle with a more representative driving schedule, shown in Fig. 4-1. The new schedule was applied to all 1972 and subsequent model year light-duty vehicles and engines. A vehicle equipped to measure velocity and acceleration was driven over a city course in Los Angeles to determine the velocity vs. time for a typical vehicle in rush-hour traffic. (A recording of the velocity vs. time measure is used to test vehicles on a chassis dynamometer, as discussed in the sidebar, 1975 Chassis Dynamometer Emission Test Procedure.)

55% of Composite Fuel Economy

➡ 31.5 Min Running Time
➡ 11.0 Miles
➡ 21.2 MPH Avg
➡   23 Stops

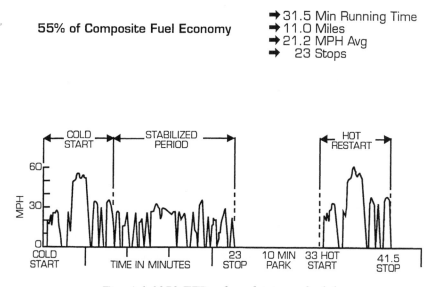

Fig. 4-1 1978 FTP urban driving schedule.

1975 Chassis Dynamometer Emission Test Procedure

The chassis dynamometer is now standard automotive test equipment used to measure emissions, because obtaining such data on a vehicle driven over a road course is prohibitively expensive. With the vehicle securely tied down and the driving wheels mounted on dynamometer rolls, the vehicle can be "driven" while not actually moving.

Prior to being tested for emissions, a vehicle must "soak" for at least 11 hours in a controlled atmosphere so that the temperature of the vehicle is between 20 and 30°C when the vehicle is pushed to the chassis dynamometer for the driving test. (A completely independent test for evaporative hydrocarbons, known as the "shed" test, measures emissions from a fully warm vehicle after it is parked in an airtight room. The vehicle must remain in the room long enough to simulate parking overnight and emit a total hydrocarbon quantity less than a specified value.)

With the dynamometer calibrated for the specific mass of the vehicle, and the vehicle at room temperature, a human operator starts the vehicle. The engine is allowed to idle for 5 seconds, then the automatic transmission is shifted into drive, and 15 seconds later the vehicle begins to "move," following a predetermined vehicle speed vs. time trace. The vehicle is "driven" approximately 7.5 miles, with 17 starts and stops, for a total of 18 cycles, one cycle between each start and stop. Vehicle top

The first "revised" driving schedule was originally established as an eighteen-cycle procedure, beginning with a cold start [4.9]. (A cycle is defined as the distance between each start and stop.) However, the auto industry registered strong complaints that not all vehicle trips originate with a cold start. So, after much deliberation, the driving schedule was revised to retain the original schedule with a cold start preceding the first eighteen cycles, but with a hot start added, followed by a repeat of the first five driving cycles. The hot start followed the original cold start and its eighteen cycles, after a ten minute "soak" (a period when the vehicle is motionless, the hood is open, and a fan blows cooling air at the front of the vehicle). Then, to retain the original legislated

speed is 56 miles per hour. To qualify as an acceptable driving test, the vehicle speed cannot deviate from the prescribed schedule by more than 2 mph for more than 2 seconds.

Emissions from the vehicle are sampled by extracting some exhaust gas from the tailpipe. Total emissions are measured during the driving schedule. The emissions output for the vehicle, in grams per mile, is calculated by dividing the emissions output in grams by miles driven.

To test the many models manufactured by General Motors (GM), a chassis dynamometer facility was constructed at the GM Proving Grounds. This facility had the capacity to test up to 36 vehicles on the Automobile Manufacturers Association (AMA) schedule, with test monitoring from a central control room. A prototype vehicle representing a model destined for release the following year had to be run for 50,000 miles with no modifications to the vehicle, except for scheduled maintenance and any unscheduled maintenance approved by EPA. Using the official driving procedure, emissions were measured at zero miles and at every 4000-mile increment until the vehicle reached 50,000 miles.

emissions, the cold-start five cycles and hot-start five cycles were combined with weighting factors. The weighting factor for emissions output was established at 43% for the cold-start eighteen cycles and 57% for the hot-restart five cycles. An EPA ruling made this driving procedure effective for the 1975 model year. It was thus abbreviated as 75 FTP (Federal Test Procedure). In the United States, the total emissions of HC, CO, and NO_X resulting from this driving procedure were the same in grams per mile, regardless of the size of the vehicle. Different vehicle sizes were accounted for by setting an appropriate inertia weight into the chassis dynamometer used to test the vehicles.

During the test, gaseous emissions are sampled by means of a probe inserted into the exhaust and are processed using a constant volume sampler (CVS). The sampled exhaust gas and a fixed quantity of background air are pumped into a dilution box. The CVS pump then draws a sample of gas from the dilution box into one of the plastic bags attached to the CVS fittings. The other bag is filled with background air. Gases from these two bags are then analyzed using NDIR (non-dispersive infrared) for CO; FID (flame ionization detectors) for HC, and chemiluminescence for NO_X, and the results are converted from parts-per-million to grams per mile.

The Clean Air Act of 1970 required light duty vehicles to meet emission control standards for 50,000 miles. The Exhaust System Task Group of the Automobile Manufacturers Association (AMA) agreed on a 59-mile test route, requiring approximately two hours. This AMA durability schedule included suburban roads, city streets, heavy traffic and some freeway driving on city and suburban streets between Dearborn and Warren, Michigan, by way of Detroit (Fig. 4-2) [4.10].

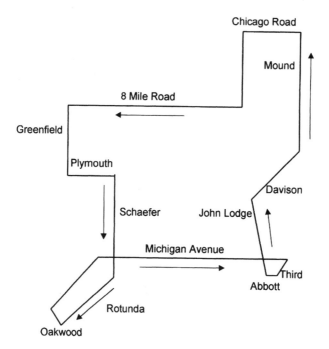

Fig. 4-2 AMA test route.

With a mix of stop-and-go operations with gentle accelerations and very little hill climbing, this schedule proved to be a "mild" driving schedule for aging an emission control system, especially since vehicles were stopped and started only for fueling and maintenance.

Vehicle manufacturers in the United States are required to deliver these prototype durability vehicles to EPA testing laboratories for an emission test at 12,000, 24,000, 36,000, and 50,000 miles for confirmation emission testing. Vehicle manufacturers from foreign countries are required to provide emissions data to EPA only after 50,000 miles of testing, without the intermediate emissions data.

Emissions data measured at 12,000, 24,000, 36,000, and 50,000 miles for U.S. prototype vehicles are used to generate a correlation of emissions vs. mileage from 0 to 50,000 miles. The ratio of the emissions value at 50,000 miles to that at 4000 miles identifies an emissions deterioration factor (DF) for each regulated constituent.

For vehicles produced in the United States, EPA approves a vehicle to be copied in a production version if the durability vehicle satisfies emission standards after 50,000 miles of durability testing. However, prior to certification of a vehicle family for sale, "data" cars must be assembled using "production-intent" components. Then, emissions from these data cars are measured after 4000 miles of durability driving. Emission data from the data vehicles at 4000 miles are extrapolated to 50,000 miles using the deterioration factors. Each data car represents a vehicle family, and if the emissions values extrapolated to 50,000 miles exceed the appropriate standards, the vehicle family cannot be produced.

Beginning in the mid 1980s, EPA allowed the automobile manufacturers to use an alternative durability procedure (ADP) based on aging a catalytic converter in an engine-dynamometer test. The engine dynamometer was operated to simulate the severity of on-road driving conditions by matching critical temperatures, exhaust flows, and exhaust gas constituents to duplicate those factors measured on a specific vehicle operated on a road aging schedule. These procedures allowed the vehicle manufacturers to test different experimental vehicles without having to age a catalyst on every vehicle, thus requiring less time to complete aging tests at significantly reduced cost.

In addition to the mileage certification process, random vehicle inspections are now used, especially in California, to determine if a vehicle family exceeds allowable emissions standards as the vehicle fleet ages. When the Clean Air Act was revised in 1990, the U.S. driving schedule was changed considerably to include emissions at higher speeds; at colder ambient temperatures; and with the air conditioning system operating. These changes are detailed in Chapter 7.

European Emissions Test Procedures

The Treaty of Rome in 1957 established the European Economic Community (EEC), and thereby initiated process of integrating the individual European countries into one community by creating a common internal market. This treaty established a working relationship between four European institutions: the Commission, the European Parliament, the Economic and Social Committee, and the Council of Ministers. Founding members of the EEC included Belgium, Germany, France, Italy, Luxembourg, and the Netherlands. Shortly afterward, Denmark, Greece, Ireland, Portugal, Spain, and the United Kingdom also joined the EEC.

The Economic Commission for Europe (ECE), was established at the 1958 Geneva Convention to implement common regulations for business and commerce. Negotiations between the member countries resulted in the ECE essentially accepting the emission control regulations established by the EEC for the automobile industry [4.11]. Reflecting the concerns of citizens in the late 1960s, both of these organizations recognized the need for effective exhaust gas regulations. Consequently, in March 1970, the EEC adopted directive 70/220/EEC followed by the UN-ECE Regulation 15 with amendment 04, which established regulations for CO and HC controls for the first European driving schedule. The overall objective of this directive was to ensure that reductions in automotive emissions in Europe would be similar to those in the U.S., taking into account different fleet composition and driving patterns. In the time period 1970 to 1983, these European emission regulations have cut HC, CO, and NO_X levels by 56%, 68%, and 32%, respectively [4.12].

The original emission levels were based on vehicle weight. However, in 1985 member states of the EEC agreed to the "Luxembourg Agreement," Directive 88/76/EEC, which defined emission limits based on engine displacement [4.13]. These regulations remain in effect today, with the result that European emission regulations are more complicated than those in the United States, that is, European regulations have separate emissions levels depending on engine displacement, which loosely correlates with vehicle size. Vehicles are divided into three classes based on engine size: (1) 1.4 L or less, (2) 1.4 to 2.0 L, and (3) 2.0 L or larger.

A European driving schedule was finally agreed upon in 1970, but only after much deliberation about what to include to represent a "typical" driving trip in Europe. The schedule that became effective in 1970 is often referred to as schedule 15-04; it includes four repeats of three cycles, shown as the first 820 seconds in Fig. 4-3 [4.14]. Emission levels from this driving procedure must meet a separate standard for CO and a combined standard for HC + NO_X, according to directive 83/351/EEC.

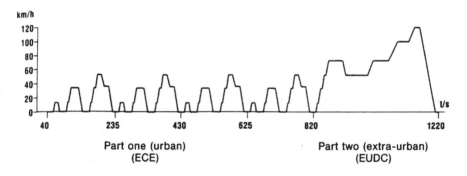

Fig. 4-3 European driving schedule (ECE+EUDC).

This driving schedule was modified in 1990 in response to concerns that it did not adequately account for high-speed driving, which is common on European autobahns. After much deliberation, a high-speed section, the "extra urban driving cycle (EUDC)," was added to the driving schedule, and EEC began phasing in emission controls based on it, EC93 and EC96 (Fig. 4-3). Different aspects of the European ECE+EUDC driving schedule are compared to those of the US 75FTP in Table 4-1.

Table 4-1. Comparison of Driving Schedules

		US 75FTP	ECE+EUDC
Actual driving time	(s)	1877	1220
Length	(km)	17.8	11.0
Average speed	(km/hr)	34.1	33.6
Maximum speed	(km/hr)	91.2	120.0

Tailpipe emission levels using this new ECE+EUDC driving schedule were established in 1990 at 2.72 g/km for CO and 0.97 g/km for $HC + NO_x$. These emission levels measured for the ECE+EUDC driving schedule may be slightly more stringent than the 1983 U.S. levels measured using the U.S. 1975 FTP [4.15]. This is because in the ECE+EUDC schedule lower vehicle speeds immediately following the cold start delay heating of the catalytic converter to light-off temperatures. In addition, the ECE+EUDC driving schedule includes vehicle operation at high speeds of 120 km/h.

The ECE+EUDC driving schedule was phased in beginning in 1993. Emissions regulations to accompany this driving schedule are outlined in Table 4-2, with levels identified as Euro I for 1993 and Euro II for 1996. Euro III levels were targeted for the year 2000 and Euro IV levels were targeted for 2005 [4.20].

Table 4-2. European Union Passenger Cars Emissions Standards

Emission Standard	EURO I EC93	EURO II EC96	EURO II EC96	EURO II EC96
		Gasoline (g/km)	Diesel, Indirect Injection (g/km)	Diesel, Direct Injection (g/km)
HC + NO$_X$	0.97	0.5	0.70	0.9
CO	2.72	2.2	1.00	1.0
PM (Diesel only)	0.14	—	0.08	0.1

Euro I regulations were adopted by directive 91/441/EEC, and Euro II by regulation 94/12/EEC [4.16]. Euro III regulations for the year 2000 include a significant modification to the ECE+EUDC driving schedule: emissions are to be measured immediately after a cold start instead of waiting 40 seconds for the vehicle to warm up at idle. To meet Euro III emissions values, targeted for 2000, European vehicles will require essentially the same technologies as those used for California low emission vehicles (LEV), as established with the U.S. Revised Clean Air Act of 1990. This act is discussed in detail in Chapter 7.

The European emissions regulations do not include a specific durability test procedure to establish compliance. Directive 88/76/EEC states that "the technical measure taken by the manufacturer must be such as to ensure that the emission of air polluting gases is efficiently limited throughout the normal life of the vehicle under normal conditions of use." This statement is interpreted to mean 80,000 km for Euro I and Euro II. For Euro III, durability requirements are targeted for 2000.

The European emissions regulations outlined above have been embraced by many European countries including Great Britain, Spain, Portugal, France, Germany, the Netherlands, Luxembourg, Switzerland, Denmark, Austria, Italy, Norway, Sweden, and Greece.

Japanese Emissions Test Procedures

The people of Japan have also recognized exhaust from automobiles as a significant source of gaseous emissions in the atmosphere. In parallel with similar events in the United States, the Japanese government has established a driving cycle, allowable emissions, and procedures for measuring gaseous emissions from automobiles. The Japanese emissions test driving cycles include an 11-mode cold start test and a 10-mode hot start test (neither of which measures cold start emissions). Vehicle speeds for these driving tests are shown in Fig. 4-4.

1. 11-mode cycle

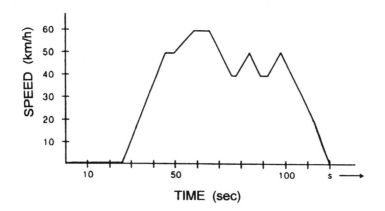

2. 10-mode cycle

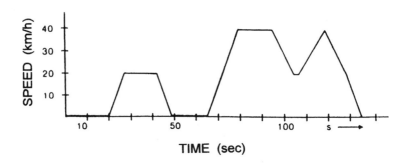

Fig. 4-4 Japanese driving schedules.

The 11-mode cold start test begins after a cold start and a 25-second idle period. The speed vs. time cycle is repeated four times, and emission measurements are recorded for all four cycles. For the 11-mode test, the vehicle is preconditioned for 15 minutes at 40 ± 2 km/h, followed by an idle period, followed by 5 more minutes at 40 ± 2 km/h. The 10-mode hot start test is repeated six times, with emission measurements recorded only during the last five repeats. Transmission gear settings are specified for manual transmissions. Automatic transmissions are specified to be in "drive" during the entire test.

Exhaust emissions are analyzed using the same basic CVS approach used in the United States. Vehicles built in Japan have been required to meet emission control standards since 1975. All Japanese vehicles shipped to the United States have always met the appropriate emissions regulations, even in U.S. driving tests.

The emission driving and test procedures described above are used in most countries throughout the world. As a consequence, vehicles manufactured after 1980 in any one country which meet that country's emissions standards will likely pass equivalent emissions tests in other countries. However, because of vehicle variability, it is still difficult to make definitive comparisons of emissions from the same model-year vehicles tested using alternative test procedures [4.17].

In many countries, including the United States, regulating agencies continue to try to convince all countries to use the same emissions testing procedures [4.18]. However, as regulations become more restrictive and tailored to individual regions of the world, it seems unlikely that this goal will be reached.

The Role of Industry

Today's automobile industry embodies outstanding scientific, engineering, and manufacturing skills, and is closely aligned with an enormous supplier community. With OEMs and suppliers coordinating their efforts to develop the various technologies necessary to control emissions, private industry has shouldered the burden of meeting mandated control standards. To do so, it has drawn upon its resources to create, design, validate, implement, fabricate, and sell the additional hardware and vehicle modifications that were necessary. To achieve the goal of reducing emissions, many automotive systems, including the engine and control systems, had to be modified. And, because fuel economy was also regulated, manufacturers also sought ways to achieve overall reduction of vehicle weight.

This technological challenge required the automotive industry to adopt a totally new perspective. New marketing strategies were necessary because emissions control could not be marketed as a feature directly associated with value. That is, the industry had to find ways around the prevailing attitude of "let the other fellow pay for it."

In response to governmental and societal pressures to lower vehicle emissions, the industry launched massive efforts on several fronts, including: (1) the recruitment of staff and facilities to study air pollution; this was because the industry did not have the necessary expertise in atmospheric physics or the medical aspects of air pollution; (2) the modification of engines and control systems; (3) the study of various emission control devices; and (4) the investigation of alternative powerplants for automobiles. One example of a pioneering industry effort was the Inter-Industry Emission Control Program, described in the section that follows.

GM Smog Chamber

As early as 1956, using a long-path infrared cell, the General Motors (GM) Research Laboratories undertook laboratory studies of the gaseous reactions that produce smog. The results of these studies identified many of the complex reactions that occur in the formation of smog, some of which have been discussed previously. In addition, the reactivities of many individual hydrocarbons were measured. Some of the hydrocarbons were found to be quite unreactive; others, however, were found to have smog-forming reactivities as much as 1000 time greater than those of the unreactive species.

The conditions used in these bench tests did not duplicate those in classic California smog—at least not to the satisfaction of GM Research—so, in 1960, a large, elaborate "smog chamber" was constructed. Designed to simulate either a static or dynamic model of the atmosphere [S4.2] smog chamber was a large stainless steel "room" equipped with 247 fluorescent bulbs to

Inter-Industry Emission Control (IIEC) Program

The Inter-Industry Emission Control (IIEC) Program, spearheaded by Ford Motor Company, was a joint effort of the auto and oil companies [4.19]. It was established by industry to accelerate the development of "emission-free" gasoline-powered vehicles. Under this program, emission levels from automobiles were to be reduced to levels below those produced in the Los Angeles Basin in the 1940s. Expressed in terms of the 1975 FTP, the goals for emissions from automobiles were 1.4, 10.3, and 1.0 g/mi for HC, CO, and NO_X, respectively.

IIEC was established in April 1967 as a three-year program with expected expenditures of $7.0 million dollars. The first phase of the program, IIEC1, was initiated shortly after the U.S. Congress passed the 1963 Clean Air Act. By 1971, IIEC1 included thirteen companies: Amoco Oil, Atlantic Richfield, Fiat, Ford Motor, Marathon Oil, Mitsubishi Motors, Mobil Oil, Nissan, Standard Oil (Ohio), Sun Oil, Toyo Kogyo, Volkswagen, and Toyota. Results from IIEC1 are discussed further in Chapter 5.

provide the equivalent of "sunlight." Exhaust gases from an engine driving a vehicle on a chassis dynamometer were ducted into this chamber; and the engine was exercised in such a manner as to replicate a typical urban driving trip, with accelerations, decelerations, steady speeds, and starts and stops. The smog chamber successfully duplicated California-type smog, as testified to by volunteers who experienced eye irritation similar to that experienced in Southern California. Using a variety of sensitive instruments, this test facility permitted researchers to monitor the composition of smog and to vary many parameters, thus arriving at a relatively complete understanding of its chemistry.

Reference

S4.2   GM Corporation, "1972 Report on Progress in Areas of Public Concern," 1972.

In 1970, the U.S. Congress passed amendments to the Clean Air Act [4.9], which lowered emission standards for 1975 model year vehicles to the following: 0.41 g/mi for HC; 3.4 g/mi for CO; and 0.40 g/mi for NO_x. These emission standards were to be valid for 50,000 miles of vehicle operation. In response to these lowered standards, IIEC2 was formed in 1974, extending IIEC1 and including the objectives of better fuel economy, good vehicle driveability, and overall vehicle and emission control durability. Results from IIEC2 are summarized in Chapter 6.

When IIEC2 was finally completed in 1977, IIEC1 and IIEC2 had spent a total of $32 million of private industry monies. A result of this monumental research effort was the establishment of a large technical base for emission control systems for vehicles. In subsequent years, much of the pioneering technology developed during IIEC1 and IIEC2 has been commercialized and applied in one form or another by the automobile industry to control emissions from vehicles.

Smog Antitrust Lawsuit

In 1954, management representatives from the United States auto industry met under the auspices of the Automobile Manufacturers Association (AMA) and launched a joint program of cooperative research and development designed to find solutions to the difficult technological problems of pollution caused by motor vehicle emissions. This joint effort caused concern among environmentalists, who anticipated an industry-wide conspiracy to deny the public access to technological findings that could be used to control emissions. Thus, a lawsuit was initiated in 1969, entitled "United States vs. Automobile Manufacturers, General Motors Corporation, Ford Motor Company, Chrysler Corporation, and American Motors Corporation," but also known as the "Smog Case." The purpose of this lawsuit was to prohibit the exchange of technical information among automotive companies. The case was brought by the United States Department of Justice under the Sherman Antitrust Act, and filed in the Los Angeles Federal District Court.

Final judgement, rendered in October 1969, prohibited U.S. automobile manufacturers from entering into any plan, program, or arrangement with any other manufacturer of motor vehicles or devices. Prohibited activities included:

General Motors elected not to be a partner in IIEC, preferring to develop emission control technologies using its own internal resources, including personnel and facilities. Chrysler also did not join the IIEC program, electing to rely on internal resources and considerable support from its supplier base. Concerns relating to antitrust laws perhaps were an additional factor in these decisions.

A major outcome of GM's decision to go it alone rather than participate in IIEC's research efforts became evident in 1975, when the catalytic converter was introduced industry-wide in the United States. Every GM vehicle was equipped with a bead-bed or pelleted catalytic converter, while Ford and Chrysler vehicles were equipped with monolith-type catalytic converters.

1. The exchange of restricted information
2. The cross-licensing of future patents on devices
3. The delay of installation of devices
4. The restriction of publicity on research and development of devices
5. The joint assessment of patents on devices held by third parties
6. The requirement that patent licenses on devices be conditioned upon the availability of like licenses to others
7. The filing, unless so requested by the agency involved, of joint statements with any governmental regulatory agency in the United States authorized to issue:
    a. Emission standards or regulations for new motor vehicles
    b. Federal motor vehicle safety standards or regulations
8. Refraining from filing individual company statements with agencies noted in Item 7

A Final Word on the Roles of Industry and Government

Air pollution is a global issue. There is a fixed quantity of life-supporting atmosphere surrounding the earth and we all need and share this common resource. Thus, the problems associated with air pollution are the responsibility of all nations. Ultimately, cooperative organizations, joint procedures, and air pollution controls should be promulgated in all foreign countries. The more developed countries should lead the way, not only because of their advanced technology, but also because they are the largest per capita contributors to air pollution. The U.S. government should support research to generate innovative developments for future transportation systems. This should include research on alternative fuels for gasoline and diesel engines, alternative propulsion systems, emission control devices, special purpose urban cars, and general-purpose vehicles. Given the increasing urbanization of the U.S., a virtually non-polluting transportation system will be required in the future.

References

4.1   Haagen-Smit, A.J., "Chemistry and Physiology of Los Angeles Smog," *Ind. Eng. Chem.*, Vol. 44, p. 1342, 1954.

4.2   Morse, R.S. (Chairman of Panel on Electrically Powered Vehicles): "Automobile and Air Pollution–Program for Progress," *SAE Journal*, April 1968, pp. 36–40.
      "Air Pollution—Problem and Risks," *SAE Journal*, May 1968, pp. 47–52.
      "Technology and Control of Automotive Air Pollution," *SAE Journal*, June 1968, pp. 42–51.
      "Roles of Industry and Government in Air Pollution Control," *SAE Journal*, July 1968, pp. 39–47.

4.3   GM Corporation, "Progress and Programs in Automotive Emissions Control," presented to the U.S. EPA, 1971.

4.4   MacGregor, J.R., "Rational Attack on SMOG, Today's Major Automotive Technical Challenge," *SAE Journal*, Vol. 74, No. 1, Jan. 1966, p. 84.

4.5   Jensen, D.A., "The Public's Role in the Automobile Exhaust Emissions Program," SAE Paper No. 660103, Society of Automotive Engineers, Warrendale Pa., 1966.

4.6   State of California, "Test Procedure and Criteria for Motor Vehicle Exhaust Emission Control (revised)," Motor Vehicle Pollution Control Board, 1964.

4.7   U.S. Congress, "Environmental Pollution—A Challenge to Science and Technology," Report of Subcommittee on Science, Research and Development to the Committee on Science and Astronautics, United States House of Representatives, 89th Congress, 2nd Session, 1966.

4.8   U.S. Government, "Control of Air Pollution from New Motor Vehicles and New Engines," Federal Register, Part II, Vol. 31, No. 61, 1966.

4.9   U.S. Government, "Control of Air Pollution from New Motor Vehicles and New Motor Vehicle Engines," Federal Register, Part II, Vol. 33, No. 136, 1970.

4.10   Homfeld, M.F., Johnson, R.S., Kolbe, W.H., "The General Motors Catalytic Converter," SAE Paper No. 486D, Society of Automotive Engineers, Warrendale Pa., 1962.

4.11   Henssler, H. and Gospage, S., "The Exhaust Emission Standards of the European Community," SAE Paper No. 871080, Society of Automotive Engineers, Warrendale Pa., 1987.

4.12   Cucchi, C. and Hublin, M., "Evolution of Emission Legislation in Europe and Impact on Technology," SAE Paper No. 890487, Society of Automotive Engineers, Warrendale Pa., 1987.

4.13   Williams, M.L., "Relating Vehicle Emission Regulations to Air Quality," *Proc. Instn. Mech. Engnrs.*, Vol. 202, No. 4, 1987.

4.14   Economic Community of Europe, "Official Journal of the ECE," C81, 1990.

4.15   Barnes, G.J., personal communication, 1991.

4.16   Haddid, O. and Greg, D.W., "Emission Control for ULEV and EC Stage III Emissions Legislations," Ricardo Consulting Engineers, 1994.

4.17   Bates, S., Brisley, R., Gagneret, P., Lox, E., Rickert, G, Searies, R., Van Houtte, S., and Zink, U., "The Attainment of Stage III Gasoline European Emission Limits Utilizing Advanced Catalyst Technology," SAE Paper No. 961897, Society of Automotive Engineers, Warrendale Pa., 1996.

4.18 Barnes, G. J. and Donohue, R. J., "A Manufacturers View of World Emissions Regulations and the Need for Harmonization of Procedures," SAE Paper No. 850391, Society of Automotive Engineers, Warrendale Pa., 1985.

4.19 McCabe, L.J. and Koel, W.J., "The Inter-Industry Emission Control Program-Eleven Years of Progress in Automotive Emissions and Fuel Economy Research," SAE SP-431, *Inter-Industry Emission Control Program 2 (IIEC-2) Progress Report No. 2*, Society of Automotive Engineers, Warrendale Pa., 1978.

4.20 Bielaczyc, P., and Merkisz, J., "Euro III and Euro IV Emissions—A Study of Cold Start and Warm Up Phases with a SI (Spark Ignition) Engine," SAE Paper No. 99-01-1073, Society of Automotive Engineers, Warrendale, Pa., 1999.

Chapter 5

U.S. Emission Controls Prior to Catalytic Converters

Beginning in the 1930s, the automobile came to occupy a significant role in U.S. society. It provided the potential for unparalleled individual mobility and freedom to travel, with the primary limitation being the existing road system. In the early years of the automobile, the major challenge to both automotive engineers and vehicle owners was keeping the vehicle running beyond the first few thousand miles of operation.

With relatively few cars on the road, air pollution from engine exhaust was not a concern; the prime consideration of manufacturers and owners alike was vehicle maintenance. Fortunately, the design and layout of most of the early automobile, especially the engine compartment, was quite simple. In fact, for vehicles built prior to the 1950s, some repairs were completed using just a pair of pliers and a piece of bailing wire.

Reflecting the continuing evolution of the automobile, including the incorporation of V-8 engines, automatic transmissions, power-assisted steering and braking systems, and air conditioning; the engine compartment of the 1965 Ford Mustang, shown in Fig. 5-1, was typical of a 1960s vintage vehicle. This vehicle was built just before emission controls began to appear on the scene. The under-hood space still provided substantial working room for access to the engine and accessories. In addition, there was still sufficient space left in the engine compartment for air circulation for cooling of the external surfaces of the engine, especially the exhaust manifolds, and to permit air flow through the radiator.

Fig. 5-1 1965 Ford Mustang.

Compared to that of the 1965 Ford Mustang, the engine compartment for the 1993 Pontiac Bonneville (Fig. 5-2) was a packaging challenge.

Fig. 5-2 1993 Pontiac Bonneville.

For styling reasons, a V-6 engine is mounted transversely to accommodate front-wheel drive and a lowered hood line. Essentially, there is no open space available in the engine compartment, making repairs difficult, and leaving very little room for air circulation to cool hot engine parts. This crowded engine compartment reflects the complex integration required to accommodate the modern vehicle powerplant and its associated systems, which include sensors and actuators—as well as other products of the "electronics revolution."

Emission Controls in the 1960s

One of the first devices developed to control vehicle emissions was the positive crankcase valve (PCV) system, used to recycle crankcase vented hydrocarbons into the engine intake. PCV systems were initially installed on 1961 model vehicles in California, a practice that was extended nationwide on 1963 models, prior to the enactment of any state or federal regulations.

Crankcases must be vented because the combustible mixture of air and unburned fuel that accumulates in them can potentially explode. Before emission controls came along, pressure buildup in the crankcase was avoided by venting accumulated unburned hydrocarbon vapors, blowby combustion

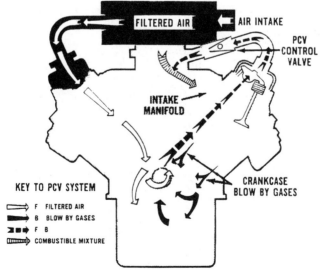

Fig. 5-3 Closed PCV system.

products, and evaporated oil into the atmosphere. Hydrocarbons vented from the crankcase accounted for as much as 20% of the hydrocarbon emissions from the 1960 vintage automobile. An additional 20% of the hydrocarbon emissions at that time were those vented from the fuel tank and carburetor. The remaining 60% of the hydrocarbons, 100% of the carbon monoxide, and 100% of the oxides of nitrogen were emitted from the tailpipe.

The PCV system ducts hydrocarbon gases from the crankcase to the intake manifold to be mixed with the fuel-air mixture for burning in the combustion chamber. Air to replace the vented crankcase gases is filtered in the induction air cleaner before being ducted to the crankcase. The PCV "check valve" permits the flow of vented crankcase gases into the intake manifold, but blocks the flow of the combustible mixture of fuel and air into the crankcase (Fig. 5-3).

Evaporative Emission Control System

(Note: This material is extracted from the reference cited below, King et al. [S5.1])

The heart of an evaporative control system is a high-grade nylon canister of approximately 3 liters volume, containing approximately 600 grams of activated charcoal, mounted in the engine compartment. The charcoal granules are supported by a porous foam screen over the bottom of the canister, which is open to atmosphere.

The activated charcoal granules range is diameter from 0.033 to 0.094 in. (0.838 mm to 2.39 mm), thus they provide approximately 170 acres (6.88×10^5 square meters) of surface and can hold 35% of their weight in liquid gasoline fuel. And because activated charcoal has a rather weak attaching force for hydrocarbon vapors, molecules of fuel adsorbed on its surface can be easily released or desorbed by flowing ambient air through the canister.

In an automobile, the primary source of hydrocarbon evaporative vapors is the fuel tank. The fuel filler cap seals the gas tank and is a one-way check valve, allowing only flow of ambient air into the tank. Tank vent ports are attached to a vent line leading to the charcoal canister.

Fuel tanks may contain as many as three venting points, so at least one vapor vent is open at all times. Orifices in these venting ports are sized to allow fuel vapor to slowly vent to equalize pressure in approximately 15 minutes.

To control unburned hydrocarbon vapors originating in the fuel tank and the carburetor, an evaporation control (ECS) system was developed prior to 1970 by vehicle manufacturers. Employing a canister containing activated charcoal, this system vents fuel vapors from the fuel tank and the fuel metering system. The activated charcoal adsorbs the hydrocarbon vapors (of reactive hydrocarbons only) before the vented gases reach the atmosphere. These adsorbed vapors are subsequently desorbed and purged into the intake manifold during the ensuing operation of the engine. Similar systems to control the discharge of evaporated fuel into the environment have been installed on passenger cars in California starting with 1970 models, and on all U.S. vehicles starting with 1971 models. Stringent legislated test methods have reduced these evaporative emissions to negligible amounts.

To prevent fuel from sloshing over vent ports during severe engine braking and other maneuvers, a vapor-liquid separator was added. To accommodate thermal expansion of liquid fuel, an air chamber was added to fuel tanks.

Another source of hydrocarbon vapors is the carburetor. Stored thermal energy in engine manifolds and heads heats the fuel metering components in the engine compartment, either a carburetor or fuel injection system, during a hot soak period immediately following shutdown of the engine. With the fuel supply system sealed, fuel evaporated during this hot soak is vented through a hose leading to the carbon canister.

Vapors adsorbed on the charcoal surfaces are purged into the engine intake system during engine operation. It is important to do this carefully so that driveability and emissions are not affected. The simplest purge system is one in which a small orifice in the tubing controls flow from the canister to the intake manifold; and the driving force pushing ambient air through the canister is the pressure depression from intake manifold. For more accurate purge control, a valve can be inserted in the purge line, which can be actuated using manifold vacuum as a pressure source, or by a solenoid responding to a signal from the electronic control module.

Reference

S5.1 King, J.B., Schneider, H.R., and Tooker, R.S., "The 1970 General Motors Emission System," SAE Paper No. 700149, Society of Automotive Engineers, Warrendale, Pa., 1970.

A significant quantity of unburned hydrocarbon vapors results when the gas volume in a depleted vehicle fuel tank is displaced by liquid fuel. Hydrocarbon vapors are also purged into the atmosphere when tanker trucks fill large underground gasoline storage tanks at filling stations (unless the filling hose is designed to recycle the vapors back into the truck). In several regions of California in the 1960s, systems were installed at service stations to recycle the displaced hydrocarbon vapors from vehicle fuel tanks into the underground fuel tank. Starting in 1999, or soon thereafter, vehicles will be required to have on-board the capacity to adsorb these displaced hydrocarbon vapors, and subsequently purge them into the engine.

In response to the Clean Air Act of 1963 and amendments enacted in 1965, automotive manufacturers in the U.S. embarked on vehicle exhaust emissions control programs to meet regulations mandated for 1966 in California and 1968 nationwide. Additional amendments to the Clean Air Act were passed by the U.S. Congress in 1970; following amendments to the Clean Air Act 1967 known as "the Muskie Bill" [5.1].

The Muskie Bill required EPA to study the cost-effectiveness of the various technologies that could be used to meet the proposed standards [5.2]. To initiate the cost-effectiveness study, the EPA, in 1971, sent a letter of inquiry to all U.S. auto companies and several foreign manufacturers. This letter requested information about proposed methods the industry might use to meet the more stringent proposed emission requirements. Thirty companies responded with cost information and information on the status of technology for candidate systems, including information on the effects of lead additives, emission levels, and durability issues [5.3]. To assess the emission control system technology, EPA representatives visited four major U.S. automobile manufacturers (the Big 3 and American Motors), three foreign automobile manufacturers, several oil companies, several catalyst manufacturers, and two major lead additive manufacturers.

The auto industry reported on a broad spectrum of emission control devices and systems that had been or were in the process of being evaluated. Evaluation methods included mathematical modeling, component bench testing, and, in most cases, prototype vehicle testing. The specific emission control devices under study included: (1) exhaust gas recirculation, (2) lean thermal reactors, (3) rich thermal reactors, (4) single bed catalytic converters, (5) dual-bed catalytic converters, and (6) three-way catalytic converters. The three-way catalytic converter was envisioned as a converter with a single bed that would simultaneously chemically oxidize HC and CO and chemically reduce NO_X.

As reported by all participants, meeting the NO_X emission level of 0.4 g/mi was going to be the biggest challenge. The three-way catalytic converter was seen as the only promising technology; and unleaded gasoline was deemed essential to prevent excessive catalyst degradation considering the 50,000-mile federal durability requirement. A cost estimate of $860 per vehicle was calculated for a system using a dual catalytic converter, a low-grade rich thermal reactor, and exhaust gas recirculation, assuming that non-leaded gasoline would be available.

Because the development of emission control technologies was in its infancy, implementation of the aggressive standards contained in the 1970 legislation for the 1975 model year were finally set at 1.5, 15, and 3.0 g/mi for HC, CO, and NO_X, respectively. After further deliberations between representatives from EPA and industry, the implementation date for more stringent standards of 0.41, and 1.0 g/mi for HC and NO_X, respectively, was postponed to 1981, and the standard of 3.4 g/mi for CO was postponed to 1983. The U.S. emission regulations and dates of enactment are summarized in Figure 6-1, on page 81, for the time period from 1970 through 1994.

As a result of these actions, for passenger cars in the U.S. 49 states, new vehicles delivered for model year 1981 had emission levels lowered 96% for HC and 76% for NO_X, compared to precontrol levels of the 1960s. In 1983, the CO had also been lowered 96%. California emissions regulations lowered HC 96%, the same as in the 49 states, but lowered CO only 92%, slightly less than in the 49 states. However in California, NO_X was lowered 83%, instead of the 76% in the 49 states, because NO_X is a direct contributor to smog.

1971 Emission Controls

In the United States, as a result of increasingly stringent regulations, emission control technologies were developed into emission control systems in a series of steps [5.4]. Several auto companies were involved in the development of emission control systems, including GM, Ford, Chrysler, and

Muskie Bill

In 1967 a very significant amendment to the Clean Air Act, known as the "Muskie Bill" [S.2], was passed by the U.S. Congress, followed closely by the 1970 Amendments to the Clean Air Act [S5.3]. These bills required HC, CO, and NO_X emissions to be lowered by at least 90% from 1970 levels. The emission standards for HC, CO, and NO_X were set at 0.41, 3.4, and 0.4 g/mi, respectively (values adjusted to the 1975 FTP). A target date of 1975/6 was set for implementing these standards.

Before specific standards were promulgated into law, this bill underwent many hearings, meetings, and deliberations among environmentalists, catalyst suppliers, automobile manufacturers, oil companies, and representatives from both the U.S. Government and California [S5.3]. EPA hosted many hearings on various subjects including particulate emissions, risk-benefit analysis of candidate emission-control technologies, and health impact estimates for CO and sulfate exposure from vehicle exhaust. One example of the huge effort expended was a 400-car fleet test conducted by General Motors at the GM Proving Ground in 1975. EPA and other auto manufacturers participated in this test, the purpose of which was to check the validity of results obtained from a computer model named HIWAY, developed by EPA to analyze and predict roadside emissions from automobiles. The results of this costly test indicated that the "HIWAY" model was not valid for all conditions. For certain meteorological conditions, it overpredicted down-wind roadside pollutant concentrations by large factors, and it could underpredict up-wind pollutant concentrations; this was because the model did not recognize the local turbulence near the exhaust pipe outlet, which was due to the turbulence created by the movement of the vehicle and the rise of the heated exhaust by convection. Based on these test results, EPA dropped its proposal for a sulfate emission standard originally targeted for 1979 [S5.3].

A further result of the deliberations surrounding the amendments to the Clean Air Act was that the emission standards for 1973 were set at interim values of 3.0, 28,

American Motors. The devices described in the following section are specifically for GM products. Ford, Chrysler, and American Motors, as well as foreign auto companies marketing automobiles in the United States, implemented similar devices.

and 3.1 g/mi for HC, CO, and NO_X, respectively (values adjusted to the 1975 FTP). The passage of this bill initiated intense research and development efforts by both the automobile manufacturers and catalyst companies. Following many hearings and sometimes heated exchanges, between the automobile manufacturers, catalyst companies, and regulators, in 1975/6 the emissions standards were reset at 1.5, 15, and 3.1 g/mi for HC, CO, and NO_X respectively. The NO_X standard was further reduced to 2.0 g/mi in 1977. As a result of ongoing negotiations, the federal standards were lowered significantly in 1981 to 0.41, 7.0, and 1.0 g/mi for HC, CO, and NO_X, respectively. In 1983 the CO standard was lowered once again to 3.4 g/mi. These federal emissions standards for passenger cars are summarized in Fig. 6-1, with all values adjusted to the 1975 FTP driving schedule and calculation procedure. California standards differed from the federal standards, usually being more stringent with regard to NO_X control and sometimes less stringent with regard to CO control, obviously because of the contribution by NO_X to ozone production [S5.4].

It wasn't until the Revised Clean Air Act was passed by the U.S. Congress in 1990 that the NO_X standard of 0.4 g/mi. was finally promulgated into law. This was phased into federal vehicle standards beginning in 1994. The phase-in was completed in 1996 (see Fig. 7-1).

References

S5.2 Lester, G.R., "The Development of Automotive Exhaust Catalysts," ACS Symposium Series No. 222, *Heterogeneous Catalysts: Selected American Histories*, American Chemical Society, 1983.

S5.3 U.S. Government, "Public Law 90-148, The Clean Air Act as amended," 1970.

S5.4 Kummer, J.T. "Catalysts for Automobile Emission Control," *Prog. Energy Combust. Sci.*, Vol. 6, pp. 177–199, Pergamon Press Ltd., Great Britain, 1979.

To meet the emission standards in 1971, the following methods were used, shown schematically in Fig. 5-4:

1. Positive crankcase ventilation (PCV)
2. Evaporation-control system (ECS)
3. Air preheat by a thermal air cleaner (THERMAC)
4. Spark control including transmission controlled spark (TCS) and a thermovacuum switch (TVS)

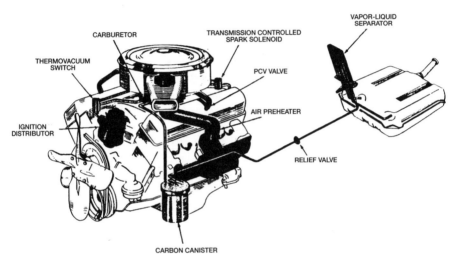

Fig.5-4 1971 emission control systems.

Preheat of Carburetor Air

During a cold start, liquid fuel droplets pass through the intake manifold and into the combustion chamber. Incomplete combustion of these fuel droplets causes low fuel economy and high emissions of HC and CO [5.5]. However, if air is preheated as it enters the carburetor, most of these unburned droplets of fuel will evaporate. Once the carburetor and intake manifold are warm enough, the fuel droplets will evaporate without being preheated.

During warm-up of the vehicle, intake air can be ducted over a portion of the exhaust manifold, and by mixing this warm air with ambient air, the air

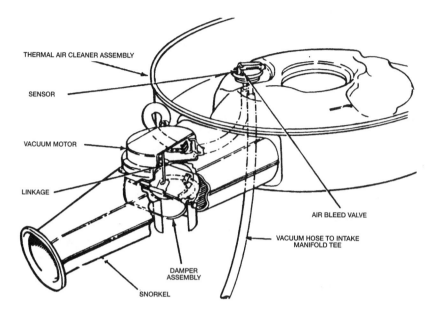

THERMAL AIR CLEANER ASSEMBLY

SENSOR

VACUUM MOTOR

LINKAGE

AIR BLEED VALVE

VACUUM HOSE TO INTAKE MANIFOLD TEE

DAMPER ASSEMBLY

SNORKEL

Fig.5-5 Thermal air cleaner (THERMAC).

temperature entering the engine can be controlled. For carbureted engines, a thermal air cleaner (THERMAC) (Fig. 5-5), includes a vacuum-operated valve to control the mixing ratio of warm and cool gases. The result is an inlet air temperature of 38°C, which allows consistent fuel metering and better control of A/F.

Control of Spark Timing

Spark timing can be adjusted to achieve many effects, such as lowered emissions, improved fuel efficiency, improved driveability, or higher exhaust temperatures to promote faster catalyst light-off. However, one spark advance setting does not optimize all desirable effects. In addition, the spark setting for best efficiency varies with engine speed and load. As shown in Fig. 5-6, spark advance varies from a minimum of 10° before top center (BTC) to 60° BTC, depending on engine speed and load.

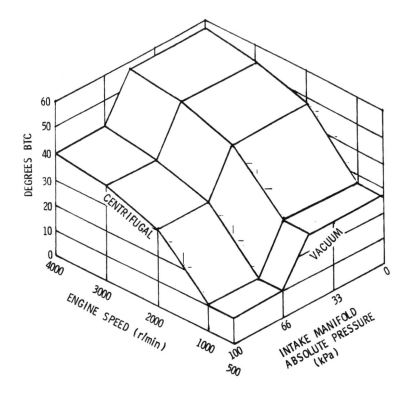

Fig. 5-6 Spark calibration for a typical 1965 vehicle.

Maximum fuel efficiency for an engine occurs at a spark setting of minimum advance for best torque (MBT). Retarding the spark setting from MBT lowers peak combustion temperature, increases exhaust gas temperature, and increases the content of unburned fuel exhausted into the exhaust manifold. These results have a significant impact on the emission control requirements.

Measured exhaust emissions from a "typical" spark-ignited engine are shown in Figs. 5-7 and 5-8. For both figures, measurements were for an engine operated at 1600 rpm with total engine throughput flow (air plus fuel) of 140 kg/h.

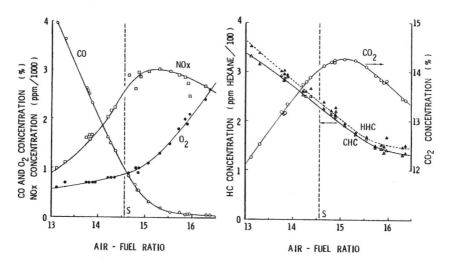

Fig. 5-7 Exhaust emissions from a spark-ignited piston engine operated at part throttle, with spark advance = 44° before top center (BTC). (Source: Ref. [5.2].)

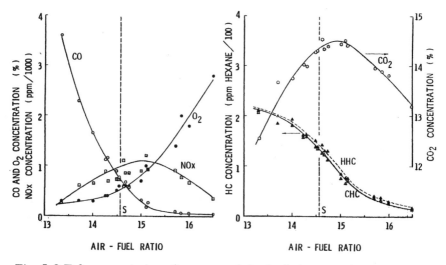

Fig. 5-8 Exhaust emissions from a spark-ignited piston engine operated at part throttle, with spark advance = 24° before top center (BTC). (Source: Ref. [5.2].)

In Figs. 5-7 and 5-8, CHC denotes hydrocarbons sampled with a cold sample line, and HHC denotes hydrocarbons sampled with a heated sample line. Because some hydrocarbon molecules are adsorbed onto the walls of a cold sample line, a heated sample consistently provides a higher value for hydrocarbons. These graphs show the same overall characteristics relative to stoichiometric A/F as described previously in Chapter 3, but include the impact of retarding the spark setting 20 degrees, that is, the significant decrease in HC and NO_X emissions generated by the engine. At stoichiometric A/F with 20° spark retard, exhaust gases may contain 30% less hydrocarbon and 60% less NO_X. Spark retard increases exhaust gas temperature, thus it can be combined with increased engine idle speed to hasten warm-up of the engine and exhaust system. However, when the spark setting is retarded, the combustion event is not completed, resulting in unburned fuel in the exhaust. Moreover, when this engine is operating at part load, fuel economy is lowered approximately 10% as a result of retarding the spark setting 20 degrees.

To meet the 1971 emission control requirements, a transmission-controlled spark (TCS) and a thermovacuum switch (TVS) were used to control spark advance. The TCS solenoid controlled emissions by blocking the vacuum advance line to the distributor, except when the transmission was shifted to high gear. The result was a fuel economy penalty during all vehicle operation in lower gear ranges. The TVS was used to reopen the vacuum line to reinstate the vacuum advance whenever retarding the spark caused engine overheating.

To lower HC and CO emissions to 1973 emission levels, additional, or secondary, air was injected into the exhaust ports upstream of the exhaust manifold to promote thermal oxidation of the CO and HC remaining in the effluent from the cylinders. These reactions can take place in the exhaust manifold if the temperature exceeds 730°C and the gases remain in contact for a sufficient residence time. On most 1973 vehicles, a belt-driven rotary-vane pump supplied this secondary air. The GM system was identified as an air-injected reactor (AIR) [5.6]. A typical 1973 emission control system is shown schematically in Fig. 5-9. In 1973, spark settings were modified and a more sophisticated system was installed to preheat air entering the carburetor to improve fuel vaporization, especially for cold ambient temperatures.

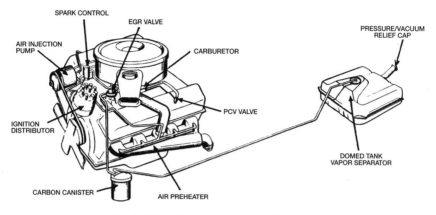

Fig. 5-9 1973 emission control system.

An alternative technique to inject additional air into the exhaust manifold, ultimately known as PULSAIR [5.7], utilized pressure pulsations in the exhaust manifold and a quick-acting check valve (Fig. 5-10). In this technique, the check valve allows air into the exhaust manifold when the pressure oscillates to subatmospheric levels, and the check valve blocks flow from the exhaust manifold to the atmosphere when the pressure inside the exhaust manifold oscillates to levels above atmospheric. Because of augmented pressure pulsation in the exhaust, PULSAIR is more effective on engines with few cylinders; it was first introduced in1975 in the four-cylinder Cosworth Vega.

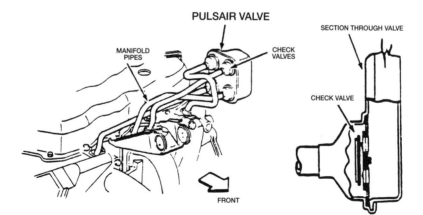

Fig. 5-10 PULSAIR.

Air Injection and Thermal Reactors

(Most of the following material is extracted from three SAE papers cited below, Steinhagen et al. [S5.5], Schweikert et al. [S5.6], and Hinton et al. [S5.7].)

Thermal reactors were popular devices with emission engineers because they were extensions of the air injection reactor (AIR) systems used by GM in the production of many vehicles beginning in 1970. Only minimal modifications to engines or exhaust manifolds were required to make AIR systems work, but developing dependable systems required considerable engineering effort.

AIR systems function by injecting additional air into the exhaust stream as it exits the exhaust passage, thus ensuring that adequate oxygen is available to oxidize all of the CO and HC present in the exhaust stream. The primary source of additional air is a belt-driven rotary pump, with the air being delivered via steel tubing to be injected into the exhaust gases at the exhaust ports. For durability and safety considerations, a diverter valve, pressure relief valve, and check valves are included in this air delivery system.

The diverter valve prevents explosions in the exhaust manifold. It does this by diverting air away from the exhaust gases at the beginning of a deceleration, when the throttle of the carburetor is closed and a momentary spike of unburned fuel is produced in the exhaust. If air is present at this moment, the unburned fuel can explode and damage exhaust components.

The pressure relief valve protects the air pump from excessive pressure by discharging air during high-speed operation of the engine. The check valves, located at the inlet to each air manifold protect the pump from possible damage by the backflow of hot exhaust.

Thermal reactors that oxidize carbon monoxide and unburned hydrocarbons were studied intensively by auto companies in the early 1970s. A thermal reactor is a chamber that replaces the conventional exhaust manifold. In it, HC and CO present are oxidized to water vapor and carbon monoxide through gas phase

---

* A thermal reactor does not control NO_X, which must be chemically reduced, not oxidized. Control of NO_X must therefore be accomplished by either control of engine air to fuel ratio (A/F) or exhaust gas recirculation (EGR).

reactions.* In a thermal reactor, gas temperatures between 650 and 760°C are adequate to oxidize HC, but a minimum temperature of 760°C is required to oxidize CO. The thermal reactor must be of certain size and configuration to withstand the high temperatures necessary and to allow for adequate residence time and confinement of the gases. Thus, size and configuration are the two challenges facing emission engineers in the design and development of thermal reactors.

Thermal reactors are classified as either rich (RTR) or lean (LTR). Rich reactors operate at vehicle A/F between 12 and 13, less than the stoichiometric value of approximately 14.7. Lean reactors operate at vehicle A/F between 15 and 16, which is greater than the stoichiometric value. Whether lean or rich, the thermal reactor must provide adequate residence time to complete gas-phase reactions. Development efforts have shown that a total volume of 1.6 to 1.8 times engine displacement is required for a thermal reactor, including insulation and high temperature housings, and this volume is to be attached directly to the engine. Making space for this much volume inside the engine compartment posed a very difficult challenge.

A lean reactor does not produce any appreciable exothermic reactions; its operation depends primarily on conservation of sensible thermal energy available in the exhaust products from the engine. This means that the reaction chamber must be mounted directly on the engine, and careful attention must be given to insulating the reactor. Another challenge is controlling fuel metering so that the engine always operates at a lean A/F. Operating a spark ignition engine at very lean A/F produces less power, and makes it prone to run rough because of combustion instability.

Compared to a lean reactor, an advantage of a rich reactor is that a large quantity of unburned fuel can be supplied to achieve high temperatures. In addition, rich A/F is desirable for occasional extra power demands for quick accelerations and hill climbing, and rich A/F has always been used for successful cold starts. The major drawback of a rich reactor is that the high temperatures possible can cause distortion, damage to the reactor, and lower vehicle fuel economy. A rich reactor also requires an air pump to supply oxygen for the combustion process. The amount of air must be modulated continuously to accurately control the combustion processes and accommodate variations in flow rate,

(cont. next page)

Air Injection and Thermal Reactors cont.

temperature, and composition of the exhaust gases supplied to the reactor; parameters that vary considerably depending on the engine power level and engine A/F.

To test emissions performance and physical durability of both lean and rich thermal reactor systems, vehicle tests programs were carried out by all automobile manufacturers, with assistance from Ethyl and Dupont. At low mileage, both lean and rich reactor systems along with modified engine controls were shown to meet emission standards of 1.5, 15, and 3.1 g/mi, for HC, CO, and NO_X, respectively. However, durability experience was demonstrated only to approximately 30,000 miles, one vehicle with an Ethyl reactor being successful only to 12,000 miles. In addition, the high temperatures encountered in service dictated the use of premium, high-cost steel alloys, such as Inconel 601, to avoid excessive oxidation. Even with premium alloys, however, reactor liners distorted causing leaks, and in some cases oxidation and distortion caused holes to be burned through metal parts. In several vehicles, abrasive metal oxide particles were ingested into engines by exhaust backflow, causing serious wear of internal engine parts. Fuel economy also suffered with reactor systems, by 7 to 10% for lean reactor systems and 10 to 12% for rich reactor systems.

The development of thermal reactors was eventually discontinued in favor of catalytic converter systems. The primary reasons for this were the poor durability experience and the fuel economy penalty.

References

S5.5 Steinhagen, W.K., Niepoth, G.W., Mick, S.H., "Design and Development of the General Motors Air Injection Reactor System," SAE Paper No. 660106, Society of Automotive Engineers, Warrendale, Pa., 1966.

S5.6 Schweikert, J.F. and Gumbleton, J.J., "Emission Control with Lean Mixtures" SAE Paper No. 760226, Society of Automotive Engineers, Warrendale, Pa., 1976.

S5.7 Hinton, M.G., Iura, T., Meltzer, J., and Somers, J.H., "Gasoline Lead Additive and Cost of Potential 1975–1976 Emission Control Systems," SAE Paper No. 730014, Society of Automotive Engineers, Warrendale, Pa., 1973.

Unlike a belt-driven pump, PULSAIR requires no useful power. In addition, compared to those for a belt-driven pump, the air supply rates for PULSAIR match more closely the requirements for oxidizing HC and CO from the engine. With a belt-driven pump, a greater amount of air, as a fraction of engine throughput, is required at low speeds than at high speeds. Thus, a belt-driven air pump must operate fast enough to supply needed air at low engine speeds, but when engine speed is increased, it supplies too much air and the excess must be vented to the atmosphere.

Exhaust Gas Recirculation (EGR)

Exhaust gas recirculation (EGR) was introduced in 1973 because the NO_X emission level of 3.1 g/mi could not be achieved with engine control technologies then available. For a number of years, exhaust gas recirculation had been under development at universities and in automobile laboratories; now it became a new technology in the emissions engineer's portfolio which could be used to control the NO_X emitted from an engine.

EGR as a technique to lower NO_X emissions from engines had been the subject of investigation as early as 1960 [5.8]. Concerns about reduced engine power, heated inlet air, and engine wear were investigated both analytically and experimentally during the 1960s [5.9, 5.10]. However, it was a comprehensive modeling analysis at the University of Wisconsin [5.11] that described the parameters that would most likely be influenced by EGR: A/F, compression ratio, percent EGR, and EGR temperature. The modeling results were unable to predict the exact effects on engine operation, but did demonstrate that EGR could substantially lower NO_X emissions.

In another important study, comprehensive experimental investigations conducted at ARCO demonstrated reductions in NO_X of up to 80%, but with attendant loss of power and impaired engine response [5.12]. Control techniques to overcome the drawbacks of EGR were suggested, including modulation of EGR to limit flow for engine maximum power, and increased spark advance to improve efficiency. The results of the Wisconsin and ARCO studies prompted the auto and fuel industries to undertake extensive follow-up studies [5.13].

Exhaust gas recirculation has proven to be a powerful technique to lower NO_X emissions. The principle behind EGR is to dilute the intake fuel and air mixture with enough inert exhaust gases that the peak temperature of the combustion event is lowered and the combustion rate is slowed. With lowered peak combustion temperatures, NO_X emissions decrease; the effect is much like the effect of spark retard, which also lowers peak combustion temperatures. In fact, spark timing can be advanced to complement EGR, thereby regaining power output from the engine and counterbalancing the lowered efficiency that results from reduced combustion temperature.

Measured exhaust NO_X emissions from a "typical" spark-ignited engine are compared in Fig. 5-11 for both EGR and spark retard. It can be seen that EGR effectively lowers NO_X emissions as a function of brake specific NO_X ($BSNO_X$). NO_X emissions can be lowered significantly with 10% EGR and drastically with 15% or more EGR. The vintage engine used in this example could not tolerate more than 15% EGR. (NOTE: $BSNO_X$ is the engine output NO_X in g/mi divided by engine output brake horsepower. Percent EGR is defined as the mass rate of recirculated exhaust gas divided by the sum of mass flow rates of air, fuel, and recirculated exhaust gas.)

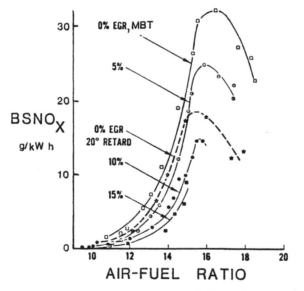

Fig. 5-11 Effect of EGR on $BSNO_X$, 5.7-L V-8 at part load.
(Source: Ref. [5.2].)

Control of EGR must be integrated with spark advance to optimize power output and fuel economy. Compared with using spark retard to lower NO_X, EGR does not incur much of a fuel economy penalty if spark advance is optimized. In fact, with spark advance selected for maximum power with EGR, as shown in Fig. 5-12, fuel economy may actually *increase* compared with operation of the engine without EGR.

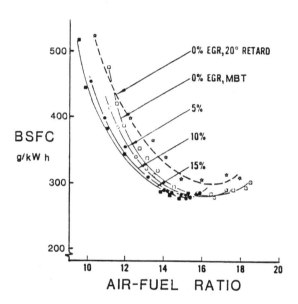

Fig. 5-12 Effect of EGR on BSFC, 5.7-L V-8 at part load.
(Source: Ref. [5.2].)

Increasing EGR dilutes the fuel/air mixture, essentially lowering the maximum power output for a given volume charge of air and fuel. In order to maintain the same power, a larger throttle opening or more engine volume flow is required. To maintain maximum power, a larger displacement engine is required. Fig. 5-13 shows the effect of EGR on engine parameters. In this figure it can be seen that spark advance approached 65° for this engine before the combustion chamber reached what is known as the "driveability limit," i.e., the point at which it is unable to support complete combustion of the fuel and air charge.

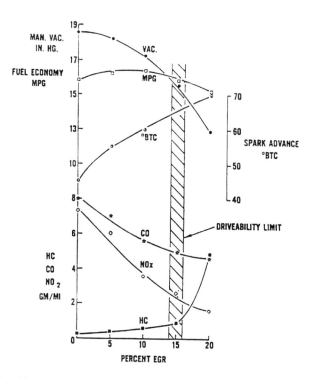

Fig. 5-13 Effect of EGR on engine parameters, 7.7-L V-8 at part load, spark advance set for best economy. (Source: Ref. [5.2].)

It can also be seen from Fig. 5-13 that the incomplete combustion associated with the driveability limit significantly increased engine-out HC emissions. Prior to reaching the driveability limit, engine-out NO_X emissions decreased sharply with increased EGR, CO emissions decreased somewhat, and HC emissions increased slightly.

Although the combustion chamber of this particular engine was unable to tolerate EGR levels in excess of 15% before the onset of unstable combustion, 20 years of development efforts have resulted in fast-burn engines that maintain complete, stable combustion at EGR levels in excess of 20% (Fig. 5-14) [5.14]. For example, stratified charge engines, in which the air and fuel mixture is stratified, provide complete, stable combustion, and tolerate EGR levels in excess of 30%.

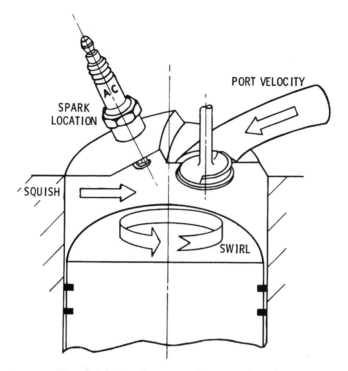

Fig. 5-14 Fast-burn combustion chamber.

For the best combination of fuel economy, lowered emissions, and driveability, EGR and spark timing must be controlled interdependently. As the load and speed of an engine are increased, a lower residual-gas fraction is trapped in the cylinder from the previous combustion event. In effect, this trapped gas is "internal EGR," so the fraction of internal EGR is lowered as engine speed and load are increased. It follows that to maintain total EGR rates, external EGR must be controlled, in part as a function of engine speed and load. To accomplish this, a valve is mounted on the intake manifold at a location adjacent to both an intake and an exhaust passage. Because exhaust pressure is higher than intake manifold pressure, by connecting the two passages with a valve, exhaust gas will flow into the intake manifold when the valve is opened. A cross section of an EGR valve is shown in Fig 5-15. Vacuum acting on a diaphragm opens the valve against a spring force. The EGR rate is controlled by controlling the vacuum signal to the valve.

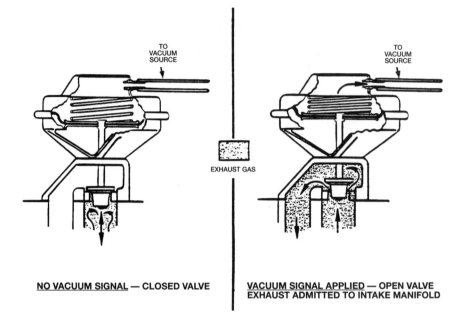

TO
VACUUM
SOURCE

TO
VACUUM
SOURCE

EXHAUST GAS

NO VACUUM SIGNAL — CLOSED VALVE

VACUUM SIGNAL APPLIED — OPEN VALVE
EXHAUST ADMITTED TO INTAKE MANIFOLD

Fig. 5-15 Vacuum modulated EGR valve.

A valve controlled by exhaust back pressure is shown in Fig. 5-16. In this design, pressure in the exhaust system acts on a diaphragm to offset the manifold signal and control EGR flow. This particular technology is now outdated. Since the introduction of computer controls on vehicles in 1980, EGR has been controlled either directly or indirectly by computer.

Please note that the emission-control systems and hardware components described above are those designed for General Motors cars, and therefore are the systems with which the author is most familiar. Ford, Chrysler, and foreign manufacturers marketing vehicles in the United States had similar development programs and produced similar emission-control systems and components.

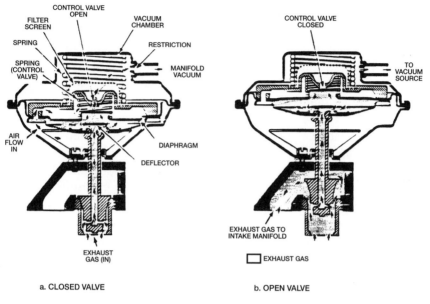

Fig. 5-16 EGR valve modulated by exhaust backpressure.

Inter-Industry Emission Control One (IIEC1)

As mentioned in previous chapters, the Inter-Industry Emission Control (IIEC) Program, established in 1967, provided the industry with much valuable technology for use in developing emission controls. IIEC1, the first phase of the program, was coordinated by Ford Motor Company in cooperation with Mobil Oil Corporation. This program encompassed at least 20 different projects, including studies of the control of HC, CO, and NO_x emissions during vehicle operation and evaporation losses. The methodologies employed included mathematical modeling, engine dynamometer testing, and vehicle testing. Technologies generated by IIEC1 studies were incorporated into prototype emission control systems before catalytic converters were introduced.

Initial emphasis was on the thermal reactor. Studies were made of physical designs that would provide high efficiency and acceptable durability. However, because only adequate durability was achieved, combined with poor driveability and fuel economy penalties, efforts were directed to catalytic converter systems.

Many EGR systems to control NO_X were designed and tested. This included studies of the interactions between EGR and spark advance. The uses of manifold vacuum and exhaust backpressure to modulate EGR were also studied. Further, in an attempt to improve fuel homogeneity to improve efficiency, lower emissions, and improve driveability, considerable research was done on fuel additives.

Reducing NO_X to very low levels was recognized as a formidable task if spark control and EGR were to be the primary emission control techniques. The use of a catalyst appeared to be potentially the best solution. However, finding the best way to control engine parameters and identifying devices that would provide a reducing gas atmosphere for NO_X and an oxidizing gas environment for HC and CO remained a dilemma. Many catalyst formulations were bench tested, including both reducing and oxidizing candidates. Several simple catalyst systems were tested on vehicles and compared with thermal reactor systems. By the end of 1973, the first phase of the IIEC program ended, and a follow-on program, IIEC2, was established. The accomplishments of IIEC2 are summarized in Chapter 6.

References

5.1   U.S. Government, Public Law 90-148, "The Clean Air Act as Amended," 1970.

5.2   U.S. Government, "Control of Air Pollution from New Motor Vehicles and New Motor Vehicle Engines," Federal Register, Part II, Vol. 33, No. 136, 1970.

5.3   Hinton, M.G., Jr., Iura, T., Meltzer, J., and Somers, J.H., "Gasoline Lead Additives and Cost Effects of Potential 1975–1976 Emission Control Systems," SAE Paper No. 730014, Society of Automotive Engineers, Warrendale, Pa., 1973.

5.4   Mondt, J.R., "An Historical Overview of Emission-Control Techniques for Spark-Ignition Engines: Part A—Prior to Catalytic Converters," ICE-Vol. 8, *History of The Internal Combustion Engine,* Book No. 100294, ASME, 1989.

5.5   Pao, H.C., "The Measurement of Fuel Evaporation in the Induction System During Warm-Up," SAE Paper No. 820409, Society of Automotive Engineers, Warrendale, Pa., 1982.

5.6 King, J.B., Schneider, H.R., and Tooker, R.S., "The 1970 General Motors Emission System," SAE Paper No. 700149, Society of Automotive Engineers, Warrendale, Pa., 1970.

5.7 Gast, R.A., "Pulsair—A Method for Exhaust-System Induction of Secondary Air for Emission Control," SAE Paper No. 750172, Society of Automotive Engineers, Warrendale, Pa., 1975.

5.8 Kopa, R.D. and Kimura, H., "Exhaust Gas Recirculation as a Method of Nitrogen Oxides Control in an Internal Combustion Engine," APCA 53rd Annual Meeting, Cincinnati, Ohio, 1960.

5.9 Kopa, R.D., Jewell, R. ., and Spangler, R.V., "Effect of Exhaust Gas Recirculation on Automotive Ring Wear," SAE Paper No. S321, Society of Automotive Engineers, Southern California Section, Los Angeles, Calif., 1962.

5.10 Robison, J.A., "Effect of Exhaust Gas Recycling on Performance and Exhaust Emissions of a Single Cylinder Engine," Technical Report PR 64-15, Ford Motor Co., 1964.

5.11 Newhall, H.K., "Control of Nitrogen Oxides by Exhaust Recirculation—A Preliminary Theoretical Study," SAE Paper No. 670495, Society of Automotive Engineers, Warrendale, Pa., 1967.

5.12 Deeter, W.F., Daigh, H.D., and Wallin Jr., O.W., "An Approach for Controlling Vehicle Emissions," SAE Paper No. 680400, Society of Automotive Engineers, Warrendale, Pa., 1968.

5.13 Gumbelton, J.J., Bolton, R.A., and Lang, H.W., "Optimizing Engine Parameters with Exhaust Gas Recirculation," SAE Paper No. 740104, Society of Automotive Engineers, Warrendale, Pa., 1974.

5.14 Amann, C.A., "The Automotive Spark-Ignition Engine—An Historical Perspective," ICE-Vol. 8, *History of the Internal Combustion Engine*, Book No. 100294, ASME, 1989.

Chapter 6

U.S. Emission Controls and the Catalytic Converter

In the early 1960s, voices in the legislatures, the public, and the press urged the industry to either develop new emission control technologies for the conventional gasoline and diesel engines or come up with an alternative powerplant for vehicles. At the same time, the U.S. government-sponsored blue-ribbon panel, chaired by Robert Morse (as discussed in Chapter 4), concluded that alternative powerplants were not being adequately studied. Types of alternative powerplants suggested for further research included battery-electric, hybrid, fuel cell, gas turbine, and steam.

Responding to these societal pressures, General Motors, Ford, Chrysler, and American Motors each began to organize technical staffs and expedite research and development efforts to develop technologies to achieve the goal of lowering vehicle emissions. The combined efforts of the auto companies, the IIEC, and the catalyst companies ultimately resulted in the development of the catalytic converter. This tale of collaboration to develop the catalytic converter, and to control exhaust emissions from vehicles, is truly a remarkable environmental success story.

Preliminary efforts to develop new technologies to control exhaust emissions concentrated on adding thermal reactors and on modifying and refining engine operation. Thermal reactors were an extension of the air injected reactor (AIR) technology, described previously in Chapter 5. As early as 1966, the AIR system proved successful in California as a technique to promote oxidation of hydrocarbons and carbon monoxide in exhaust manifolds. By adding additional volume to the exhaust manifolds, and supplying enough air and unburned fuel, thermal reactors proved promising. However, a requirement of such systems was that temperatures in the exhaust manifold had to be maintained above 750°C for a sufficient residence time [6.1]. The fuel was supplied by rich operation of the engine, and additional air to maintain an oxidizing environment was supplied from an air pump.

EPA emissions standards proposed for 1975 required high conversion efficiencies, and, as discovered through thermal reactor development efforts, this meant that high temperatures of 800 to1000°C would be necessary.

GM Progress of Power

In the mid-1960s General Motors embarked on a program to build and test prototype vehicles that would demonstrate the technical feasibility of alternative powerplants. This effort culminated in The Progress of Power exhibit, a showcase of 26 special experimental vehicles, which took place at the GM Technical Center in Warren, Michigan, on May 7 and 8, 1969.

Alternative powerplants and systems featured included electric, gasoline-electric hybrid, and an intake-valve-throttled piston engine. Futuristic powerplants included Stirling, free piston, gas turbine, steam, electric, intake-valve- throttled piston, gasoline-electric hybrid, and Stirling-electric hybrid. Highlighting the exhibit were two steam-powered vehicles, the SE-124 and SE-101, which were built and tested because legislators, the technical community, and the press were touting the potential advantages of steam power for low emissions and high fuel economy.

Besides focusing on alternative powerplants, the Progress of Power exhibit included a number of full-size and small commuter vehicles powered by conventional internal-combustion engines featuring such technological advances as exhaust manifold reactors, catalytic converters, ram air supercharging, direct-cylinder injection, and alternative-fueled engines using ammonia and liquid petroleum gas (LPG).

However, temperatures in this range were found to cause thermal stresses, thermal distortions, and weakened materials, resulting in poor   durability of exhaust components. In addition, the rich operating conditions required to supply extra fuel to the thermal reactors caused fuel economy to suffer. Confronted by these findings, most development efforts were redirected to focus on catalytic converter systems.

Chronology of the Development of Catalytic Converters for Automobiles

As discussed in Chapter 5, amendments to the 1970 Clean Air Act, estab-lished very strict targets for emissions from automobiles. This legislation prompted a series of vigorous hearings and debates,   as well as a concerted effort by both auto and catalyst companies to develop technology that could be used to substantially lower emission levels. As shown in Fig. 6-1, in 1975 allowable HC emissions were to be lowered from 3.0 to 1.5 g/mi, and allow-able CO emissions from 28 to 15 g/mi [6.2].

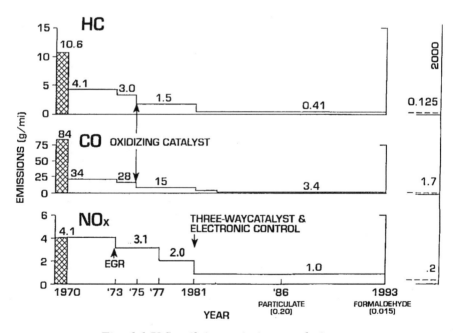

Fig. 6-1 U.S. tailpipe emission regulations.

The Magic of Steam

Steam power has always fascinated individuals involved in power generation. In fact, steam power provided the basis for the United States' progression into the industrial age. As pioneers settled the western United States, steam locomotives provided the transportation that supported   business and commerce throughout the country.

Years ago, steam power was used to power automobiles. In 1900, 40% of the cars produced in the United States were powered by steam. Well-known companies producing steam cars were White, Stanley, and Doble. In the late 1920s, General Motors even experimented with a steam-powered bus [S6.1]. However, as the internal combustion engine emerged as the engine of choice to power automobiles, steam-powered vehicles faded from the scene. Doble produced the last commercial steam car in 1929, and during the ensuing 40 years, the auto industry expended very little development effort in this direction.

During the 1960s, legislatures, educators, and the public began promoting the investigation of alternative engines to power cars. The Morse report, published in 1968, strongly suggested that additional efforts should be devoted to developing electric and steam engines as low-polluting alternative powerplants for cars. In response to growing pressures, General Motors, in 1968 and 1969, built and tested two steam-powered experimental vehicles: the SE124 and the SE101.

The SE124 was a 1969 Chevelle, selected to represent a low-powered intermediate sedan. It was redesigned with a steam engine built by Besler Developments, Inc., of Oakland, California, under contract with General Motors. At the same time, GM Research had developed a new steam engine for the SE101, which incorporated the latest technology available in 1969. The SE101 was a 1969 Pontiac Grand Prix, chosen to represent an intermediate sport coupe class of vehicle.

Both the SE124 and the SE101 were designed and built to investigate the feasibility and practicality of using steam power for modern passenger cars. The SE101 included customer convenience items such as power steering, power brakes, automatic transmissions, and air conditioning. The SE124 retained a manual transmission [S6.2] as well. Both cars were designed to condense the steam exhausted from the expander so that the vehicle operator would have to add only a minimum of make-up water.

After vehicle development progressed to the operational phase, exhaust emissions, fuel economy, acceleration, and water consumption were measured for both vehicles [S6.3]. For the SE101, fuel economy for road-load operation peaked at 7 miles per gallon at 30 mph, and decreased to 5 miles per gallon at 60 mph. For the SE124, fuel economy for road-load operation peaked at 13 miles per gallon at 30 mph, and decreased to 8 miles per gallon at 60 mph. Emissions were measured for a start-up period and on the original 7-mode HEW driving schedule. The SE101 required a 2.4-minute warm-up time and the SE124 required a 5.8-minute warm-up time. Emissions for these two operations are summarized in Table S1.

Table S1. Tailpipe Emissions for the SE101 and SE124 Steam Cars

	Hydrocarbons	Carbon Monoxide	Oxides of Nitrogen
SE101Cold Start (g)	0.7	10.3	1.6
SE124 Cold Start (g)	1.1	8.7	5.7
SE101, HEW (g/mi.)	0.5	3.1	2.1
SE124, HEW (g/mi.)	0.3	1.0	1.7

The following advantages were reported for these two steam-powered vehicles: 1) low air pollutant outputs, 2) low noise levels, and 3) good engine torque. The following disadvantages were reported: 1) large powerplant size and weight, 2) high cost, 3) water consumption, 4) potential to freezing, and 5) difficult to lubricate for long life.

When testing was completed, both vehicles were retired, their disadvantages clearly outweighing their advantages. However, they remain a benchmark challenge for any organization desiring to market a steam-powered vehicle in the future.

References

S6.1 Vickers, P.T., Amann C.A., Mitchell, H.R., and Cornelius, W., "The Design Features of the GM SE-101—A Vapor-Cycle Powerplant," SAE Paper No. 700163, Society of Automotive Engineers, Warrendale, Pa., 1970.

S6.2 Vickers, P.T., Mondt, J.R., Haverdink, W.H., and Wade, W.R., "General Motors' Steam Powered Passenger Cars—Emissions, Fuel Economy and Performance," SAE Paper No. 700670, Society of Automotive Engineers, Warrendale, Pa., 1970.

S6.3 GM Research Laboratories, "GM Progress of Power," General Motors Corp., 1969.

Techniques used to achieve reductions in emission levels prior to 1975 were studied exhaustively by all U.S. car manufacturers. It was concluded that even an optimum combination of all of these techniques would be inadequate to accomplish the required substantial lowering of allowable emissions.

In the early 1950s, the auto industry had begun development work toward installing catalytic converters on vehicles. The catalytic converter was an attractive emissions control option: It operated with no moving parts, no electric energy inputs, and no requirements for mechanical interface with the operation of the vehicle, except for a space in which to mount the device. The catalytic converter treated the exhaust gas after completion of engine combustion; it did not interfere with engine operation, and thus achieved the best efficiency, as long as added backpressure on the engine was controlled to an acceptably low level. Of all the alternatives, the catalytic converter clearly offered the best choice for emission control efficiency, durability, low maintenance, fuel economy, driveability, and cost to the consumer.

What is a Catalyst for Vehicle Emission Controls?

A catalyst is a substance that either accelerates or retards the velocity of a chemical reaction, usually by lowering the temperature at which the reaction takes place. A catalyst remains unchanged by the chemical reaction. Catalysts have been used for many years in the chemical and petrochemical industries to control reaction speeds.

Catalysts for controlling emissions from vehicles promote the oxidation of HC and CO to CO_2 and H_2O, or the reduction of NO_X to N_2. The catalysts cause these reactions to proceed at threshold temperatures of 250 to 300°C, much lower than the 720 to 750°C required for gas phase reactions. For best results, catalyst elements such as platinum, palladium, or rhodium are dispersed on a porous surface layer, such as alumina (Al_2O_3), with a myriad of tiny pores creating a very large surface area to be exposed to the exhaust gases. The surface layer may be on a pelleted substrate or in the form of a washcoat deposited on the wall of a monolith substrate (as in a monolith catalytic converter).Catalysts age by a process known as sintering. This process is

In 1959, the General Motors Research Laboratories [6.3] reported on development efforts to incorporate catalytic converters into vehicle exhaust systems. An experimental catalytic converter supplied by Oxy-Catalyst was installed on a vehicle powered by a V-8 engine. This converter was 1.01 meters (40 in.) long, 0.30 meters (12 in.) wide, and 0.20 meters (8 in.) high, and contained 11.3 kg (25 lb) of catalyst pellets [6.4]. When fresh, with additional air added to the exhaust stream to ensure an oxidizing gas stream, this oxidizing converter lowered hydrocarbon and carbon monoxide emissions by 70 to 80%. After 5000 miles of vehicle road operation using the AMA Test Route (discussed in Chapter 4) the conversion efficiencies for both hydrocarbons and carbon monoxide were lowered to approximately 30%.

However, after the same 5000 miles of durability testing, the catalytic converters showed evidence of thermal warping and distortion. The housings vibrated, making noise, and thermal distortion restricted exhaust flow. The result was unacceptable exhaust backpressure, which lowered engine power at high speeds. To protect the converter form overheating, a system to

accelerated by overheating the substrate, which collapses the pores, thus occluding the active catalytic sites. Catalyst effectiveness diminishes as a result of the aging process.

References

S6.4 Daintith, J., (Editor), *A Concise Dictionary of Chemistry*, Oxford University Press, Walton Street, Oxford OX DP, New Edition, New York, N.Y., 1990.

S6.5 Kummer, J.T. , "Catalysts for Automobile Emission Control," *Prog. Energy Combust. Sci.*, Vol. 6, pp. 177–199, Pergamon Press Ltd., Great Britain, 1979.

S6.6 Taylor, K. "Automobile Catalytic Converters," Springer-Verlag, Berlin, 1984.

S6.7 Maxwell, B. and Powell, R., "Automobile Exhaust Catalysis," *Ceramic Engineering and Science Proceedings*, 9th Automotive Material Conference, American Ceramic Society, 1980.

S6.8 Komarmy, J.M. and Klimisch, R.L., "Pelleted Catalysts for Automotive Emission Control," SAE Report No. 741049, Society of Automotive Engineers, Warrendale, Pa. 1974.

bypass exhaust was incorporated into the converter design. A valve installed at the inlet to the converter was designed to route exhaust gas around it when a thermoswitch detected temperatures exceeding approximately 700°C. In response to this bypass modification, EPA subsequently passed regulations prohibiting, as a source of excessive exhaust emissions, any bypass of exhaust past a catalytic converter. But the impact of this regulation was minimal, because researchers ultimately found that the use of a bypass valve to protect the converter was ineffective.

A misfiring spark plug is probably the most common engine malfunction that causes converter overheating. Misfiring can result from a contaminated spark plug, a worn-out spark plug, or a failure in the wiring or distributor system that delivers electric energy to the spark plug. Another malfunction that causes converter overheating is when the fueling system supplies excessive fuel to the engine. In such a situation, even though the engine may be firing properly, excess fuel is discharged into the exhaust stream. The converter oxidizes the fuel, and releases thermal energy inside the converter housing. The problem is that the thermal response of a converter is too fast for a thermal sensor and valve to react. From these examples, it is obvious that one of the major tasks facing developers of emission controls was eliminating such engine malfunctions so that a catalytic converter could be included in the exhaust system.

1966 California Emissions Requirements

In 1960, California created the California Motor Vehicle Control Board (MVPCB). Its purpose was to certify devices designed to lower undesirable emissions from automobiles, originally HC and CO, in ambient air by approximately 70% and 57%, respectively. Percentages were selected to reflect the baseline emission levels prevalent in the 1940s. The levels selected, 275 ppm HC and 1.5 volume % CO, were measured employing a vehicle operated on a chassis dynamometer, following a "7-mode" driving schedule. Nondispersive infrared instruments were specified for measuring HC and CO emissions, along with a calculation procedure for combining cold and hot emission measurements [6.5]. Control of emissions was required to be effective for 12,000 miles and had to be demonstrated on a fleet of 30 production vehicles, operated on fuel containing approximately

3 grams per gallon of lead. No modifications were allowed to the basic engine and vehicle.

In addition to the automobile manufacturers, many companies with prior industrial experience in chemistry, catalysis, and high-temperature materials responded to California's emission control challenge. In 1964, four systems designed by non-automotive companies were certified by the MVPCB for durability testing. These systems were: (1) a catalyst system by American Cynamid, and Walker Manufacturing, (2) a catalyst system by W.R. Grace and Norris Thermador, (3) a catalyst system by Universal Oil Products and Arvin Industries, and (4) an exhaust reactor system by American Machine and Foundry and Chromalloy.

At the beginning stage of the durability testing the four systems lowered emissions substantially below the target standards; however, after 8000 miles of on-road operation, none of the systems met the target standards [6.5]. Nonetheless, vehicle testing continued to the 12,000 mile durability requirement, even though the measured emissions remained significantly higher than the target values. When it became evident that the original goal of less than 275 ppm of HC was unattainable using these devices, an "average" was adopted by the MVPCB. Subsequently calculation procedures were modified to weigh emissions from hot and cold operation, so that the final emission levels met the target standards [6.6].

For the 1966 model year, California further refined emission regulations, requiring that $65 be the maximum price for an "add-on" device, either catalytic converter or afterburner, and mandating cross licensing availability, an annual maintenance cost not to exceed $15, and a useful life of 50,000 miles. From the durability data generated by this program, it was concluded that meeting the 50,000 mile requirement would dictate that available add-on devices would have to be replaced several times along the way.

During this time period, the automobile manufacturers continued their own development work on catalytic converters, at the same time evaluating converters and catalysts supplied by other companies. In addition, they diligently explored the feasibility of using emission control technologies other than the catalytic converter or afterburner. Their primary focus was on engine modifications, combined with the addition of air to the exhaust, and extensive efforts to develop exhaust thermal reactors (see earlier sections of this chapter). The auto manufacturers submitted their emission control systems to the MVPCB for approval, and, in July 1965, the following combinations of non-catalytic devices were approved: positive crankcase ventilation (PCV); evaporation-control system (ECS); air preheat by a thermal air cleaner (THERMAC); and spark control, including transmission controlled spark (TCS) and a thermovacuum switch (TVS). (Descriptions of these devices can be found in Chapter 5, under the heading "1971 Emission Controls"). The acceptance of these systems demonstrated that the auto companies could meet the California standards without having to resort to "add-on" devices [6.6].

General Motors Catalytic Converter Task Force

In 1969, under the leadership of President Edward Cole, GM assembled an interdisciplinary technical team, the mission of which was to accelerate studies of advanced systems and technologies to lower tailpipe emissions. Team members included staff from GM Research Laboratories, Engineering, Buick Division, and AC Spark Plug Division. Subsequently, a Catalytic Converter Task Force was organized to evaluate the potential of catalytic converters for automobiles. AC Spark Plug was selected as the lead component supplier division to develop and supply the catalytic converter, and Buick was selected as the car division to adapt the catalytic converter to an automobile.

The activities of the Task Force were closely monitored by GM management, and were reviewed on a weekly basis by an executive committee. Catalyst suppliers were contacted, samples of materials were obtained, and proposals by suppliers and GM personnel were carefully evaluated. As studies progressed, both mathematical modeling and laboratory experiments were used to evaluate alternative systems and catalytic converter designs. Through these studies, the task force quickly learned several important things. One was that there were very few catalysts available for use in a catalytic converter for automobiles.

Platinum, an oxidizing catalyst, was the only catalyst readily available for use in controlling automotive pollutants. The most prevalent catalyst support was a pelleted alumina, which was very porous, enabling the catalyst elements to be

---

It was a considerable relief to the auto industry to find that, in this initial compliance phase in California, they did not have to employ "add-on" devices after all. Their reluctance to employ such devices was less due to first cost concerns than their not wanting to sell and warrant vehicles equipped with expensive devices made by third parties. In this case, the devices were new to automobiles and came with the risk of potential safety and operational problems. Consequently, no add-on devices were used in 1966 model year production vehicles in California. This was a temporary setback for the catalyst companies who had spent millions of dollars in developing and promoting their devices for the market.

---

deposited into the micropores and macropores at the surface of the pellets. Such pelleted catalysts were already used extensively by oil refineries, petrochemical industries, and chemical processing plants.

Monolith catalysts were also available, for example, one that used platinum deposited on the wall surfaces of a continuous channel monolith made from mullite in a "laid-up structure," i.e., corrugated sheets with flat separators. This catalyst was used to oxidize unburned hydrocarbons and carbon monoxide from the fork-lift trucks used in mines. These early monoliths were made by American Lava Co., which was later purchased by the 3M Corporation. The first monoliths with triangular-shaped flow passages were extruded cordierite.

Another thing the task force learned was that high temperatures could easily destroy both a catalyst and a catalytic converter. Thus, if the converter component was to survive in the exhaust from an internal combustion engine, engine controls would have to be designed to limit catalyst maximum temperatures. They also confirmed the obvious: that a reducing gas mixture would be required to reduce oxides of nitrogen, and that an oxidizing gas mixture would be required to oxidize both carbon monoxide and unburned hydrocarbons.

Lastly, in a finding that would have a great impact on the oil industry, the task force identified the lead present in gasoline as a poison that could quickly render a catalyst useless. In order for a catalyst to operate for any significant length of time, lead would have to be eliminated from automotive fuels.

As a result of the efforts of the automobile manufacturers and the catalyst companies, the 1966 California emission control standards were established. This was a significant milestone in the evolution of emission control systems for automobiles. The knowledge gained and technological advances achieved by both the automobile manufacturers and the catalyst companies in meeting the 1966 California standards would prove essential in meeting the stricter emissions requirements that would follow. Notwithstanding California's initial ruling concerning "add-on devices," research had demonstrated that catalytic converters had considerable promise as a technology to lower tailpipe emissions. And it was clear that to be successful, the catalytic converter could not function alone as an add-on device, but must be incorporated as a component in an overall emission control system. So, to develop such a system would require the coordinated efforts of both vehicle manufacturers and catalyst companies.

The prototype catalyst systems built and tested by General Motors, Ford, Chrysler, and American Motors in the 1960s came in a variety of configurations. Catalytic converters were tested mounted close to the engine, in the underfloor location, and in combinations of close-mounted and underfloor locations [6.7, 6.8]. Catalyst companies supplied samples of candidate materials to be tested, including both pelleted and monolithic substrate type products.

The pelleted catalysts, containing either spherical pellets or extruded truncated cylindrical pellets, had been used successfully for many years in production in the chemical and petrochemical industries. This history suggested that the cost of using a pelleted substrate to provide a given amount of conversion would be less than that of a monolith substrate. On the other hand, the monolithic substrate offered the potential of a smaller overall size, and a shape that would be easier to fit into a small space on a vehicle, but very meager durability experience.

Studies of catalytic converters installed on automobiles revealed that to prevent a catalytic converter from deteriorating, the engine must operate without supplying excess unburned fuel. Assuming that oxygen is also available, excess unburned fuel supplied to a catalytic converter will "burn" and generate excessive temperatures. Studies also revealed the obvious, that is, that a reducing gas mixture was required to reduce oxides of nitrogen and an oxidizing gas mixture was required to oxidize carbon monoxide and unburned hydrocarbons.

To accommodate the catalytic converter, the carburetor fuel metering process was improved by designing quick-acting automatic chokes and altitude compensation to control A/F mixtures more precisely. An electronic ignition system was also introduced to provide a higher-energy and longer-duration spark discharge to improve reliability of ignition of the A/F mixture in the combustion space. Further, air injection pumps were incorporated into many emission control systems to inject additional air into the exhaust ports; the purpose was to ensure an adequate supply of oxygen, necessary for oxidation to take place in the catalytic converter, especially during engine starts. Testing and measuring the effects of these modifications produced volumes of data, and, as the results were analyzed, it became obvious to most of the automotive industry that the catalytic converter was the engineering "break-through" needed to meet the emissions regulations established for 1975/6.

Inter-Industry Emission Control Two (IIEC2)

The Clean Air Act Amendments of 1970 required target emission levels of 1.4 g/mi, 3.4 g/mi, and 0.41 g/mi for HC, CO, and NO_X, respectively. These levels were initially targeted for 1975/6. Passage of this legislation by the U.S. Congress prompted the formation of Inter-Industry Emission Control Two (IIEC2) in January 1974 [6.9]. Composed of Ford Motor Company, five oil companies, and two foreign car companies, IIEC2 carried on the work started by IIEC1.

The objectives of IIEC2 were to meet the 1975 emission targets while retaining good fuel economy, good driveability, and good durability. Specific projects focused on fuel optimization, engine design optimization, engine A/F control, and catalytic converters. Promising technologies studied and demonstrated on vehicle prototypes included dual-bed catalytic converters; EGR; and catalytic converters for simultaneous conversion of HC, CO, and NO_X. Studies of the ratio of hydrogen to carbon in fuels were focused on fuel economy and emissions impacts. Several alternative fuels and fuel mixtures were investigated in relation to cold start, octane requirement, volatility, and vehicle driveability. These included methanol and methanol-gasoline blends.

Under IIEC2, engine studies were done to map the effects of compression ratio, spark timing, A/F ratio, EGR, and combustion chamber designs in order to compare emissions, fuel economy, and octane requirements. The results of these mapping studies provided a base for establishing engine calibrations to optimize fuel economy and emissions. Stratified-charge engines and pre-chamber and open-chamber direct fuel injection engines were also built and tested.

Once the catalytic converter had emerged as the *key* element in the emission control system for automobiles, catalyst research under IIEC2 was expanded to evaluate many NO_X reduction catalysts, as well as many HC and CO oxidizing catalysts. The IIEC, and the auto companies tested both pelleted and monolithic substrates and evaluated catalyst submissions from any and every source. Extensive testing was necessary since there were few proven catalyst elements for oxidizing reactions and no catalyst elements for reducing reactions.

Bench tests, engine dynamometer tests, and vehicle tests were employed to evaluate emission performance, including such aspects as ammonia formation, catalyst durability, and the impact of washcoat "promoters" and "stabilizers" on conversion performance and catalyst durability. (Washcoats are the porous ceramic coatings attached to substrate surfaces to retain catalyst elements. Promoters enhance the catalytic reaction. Stabilizers are chemical substances used to enhance conversion performance of a catalyst and inhibit deterioration of washcoat properties.)

Developments from several of the IIEC projects were combined into an IIEC concept car using a Ford 4000 lb vehicle powered by a 351 CID V-8 engine.*

---

* The ongoing results from IIEC2 were reported as collections of papers in four SAE Publications: SP-395, 1975; SP-403, 1976; SP-414, 1977; and SP-431, 1978.

This advanced concept vehicle was a futuristic example containing most of the technologies ultimately used throughout the industry at the end of the twentieth century: a three-way catalyst mounted on each bank of the V-8 engine, an oxygen sensor, and electronic controlled fuel injection system. The on-board digital microprocessor also controlled spark timing, EGR, and secondary air. Other prototype vehicles were built and catalytic converters tested for millions of miles by both auto companies and catalyst suppliers [6.10]. General Motors alone tested prototype catalytic converters on more than 1000 cars over 25 million miles prior to releasing catalytic converters for production.

It was found that excessive temperatures generated inside catalytic converters caused both monolithic and pelleted substrates to suffer severe performance degradation, if not outright melting failures. The monolithic substrate proved more vulnerable to failure by overheating than the pelleted substrate because of a low mass per unit of surface area and the resulting quick thermal response. Obviously, converters mounted directly on the exhaust manifolds were more likely to suffer thermal damage because of the higher exhaust gas temperatures entering the catalyst. To solve the problem of catalytic converters overheating required extensive engineering effort and resulted in significant changes in the operation and control of automotive engines.

Catalysts and Lead in Gasoline

A major finding from early experimental studies of catalytic converters was that the performance of catalysts was severely diminished by lead content in leaded gasoline. Subsequent investigations revealed that conversion performance of catalysts also deteriorated as a result of phosphorous and sulfur present in gasoline and lubricating oils. However, the lead contamination of catalytic converters proved so severe that it was clear that for the use of catalytic converters to ever be practical, lead would have to be removed from gasoline. This finding generated much consternation among oil companies, who used lead as an additive to increase the octane of gasoline.

Edward Cole, President of GM, on behalf of the industry, took on the monumental task of convincing the oil companies that the lead had to be removed from gasoline to accommodate the use of catalytic converters on automobiles. This "demand" imposed on the oil industry was initiated at a meeting of the American Petroleum Institute (API) in 1970. After much dialogue, the U.S. Congress passed legislation that phased out the use of lead in gasoline fuels. Unleaded fuel, containing less than 0.03 g/gal instead of the 2.0 g/gal it contained before, became available nationwide in the U.S. in1972. Meanwhile, to enable vehicles to operate on unleaded fuel, the auto industry designed engines with lowered compression ratios and upgraded valve seats. Beginning on January 1, 1997, all gasoline fuel sold in the United States could not contain a lead content greater than 0.03 g/gal.

Making Substrates

Pelleted Substrates

The pelleted substrates used to support catalyst elements are made from aluminum oxide (Al_2O_3). To make spherical pellets, alumina powder is mixed with a solvent (predominantly water) to produce a slurry having the consistency of pancake batter. This slurry is deposited onto a horizontal plate containing thousands of holes, and the plate is mounted at the top of a high tower. The slurry is allowed to drip through the holes and free-fall through air to the base of the tower. As the slurry droplets fall, surface tension causes them to take on a spherical shape. The newly formed spheres are heated in an oven to remove the remaining water vapor, resulting in aluminum oxide pellets that are somewhat porous.

Another procedure used to make spherical pellets is to deposit powdered alumina at the center of an inclined rotating disc while spraying small water droplets onto the disc. The water causes the alumina to collect, forming small spheres, which continue to grow until they are large enough that centrifugal force causes them to fall off the disc. These pellets are then heated to remove the water vapor and solidify them.

Pellets may also be extruded as truncated cylinders. To make extruded pellets, a slurry of aluminum oxide is forced by a ram through a horizontal plate having many holes; each hole being approximately 3.2 mm (0.125 in.) in diameter. As the slurry exits the plate, the cylindrical extrudates are cut or break at a length of approximately 3.2 mm (0.125 in.). These cylindrical extrudates are solidified in

Dual Converter Systems

A major drawback of the oxidizing catalytic converter, the primary system in use from 1975 until 1980, was its inability to reduce oxides of nitrogen. To overcome this limitation, a "dual" converter system, with a reducing catalyst mounted upstream of the oxidizing catalyst, emerged as a practical solution. With this arrangement, the engine fuel metering was set slightly rich to provide the reducing exhaust gas to the first catalyst, with minimum penalty in fuel economy. Then, secondary air from an air pump was injected after the upstream converter and before the downstream converter, to provide oxygen for oxidizing unburned hydrocarbons and carbon monoxide.

a liquid bath and are subsequently heated in an oven to remove the water and further solidify the pellets.

Monolithic Substrates

To make a monolithic substrate, a ceramic powder, cordierite is mixed with a solvent (predominantly water) to produce a slurry having the consistency of pancake batter. This slurry is extruded through many tiny holes machined in a very intricate die. The holes in the die on the upstream face are located at the apexes of the extruded cross-section geometry. For example, for a monolith with square passages, the holes are located at the intersections of the four webs of the square. The die is fabricated so that the inlet hole changes shape as it passes through the die. This means that the outlet face represents the final form of the monolith, and that the extruded material must "knit" or stick to the adjoining material at the center of the span of a monolith passage—obviously a challenging processing operation.

The overall diameter of an extruded monolith is the diameter of the final element, or "brick" as it is sometimes called. This extrudate is likened to an "elephant tusk" and must be cut into finite lengths, corresponding to the flow length of a completed monolith brick. After the extrudate has been cut into final lengths, the bricks are calcined, or heated, in an oven to complete the formation of the cordierite and remove excess water. Cordierite is not porous like alumina, so to support the catalyst elements, a "washcoat," i.e., a surface layer of porous material, must be attached to the walls of the monolith channels. This washcoat is generally either a derivative of aluminum oxide or a material with similar properties.

An "integrated" dual converter system built into one housing was also developed; it provided a compact arrangement to accomplish the dual role of oxidizing and reducing, with air injected between converter elements in one housing. The concept of "dual" converters has since been refined, and a variety of configurations have been employed, some with more than two converters. Most multiple converter configurations utilize separate housings for separate converters. Multiple converter systems are usually associated with larger vehicles and larger engines.

Catalyst Elements

Platinum and palladium are the primary noble metal catalysts used to promote the oxidation of HC and CO in the exhaust from automobile engines. In catalytic converters, these noble metals are very finely dispersed on either alumina pellets or cordierite extruded monoliths. More expensive noble metal catalysts, rather than base metal catalysts, are used because of their superior catalytic activity and resistance to deterioration.

Alternative Catalysts Elements

Many base metals such as copper, chromium, manganese, iron, and nickel were investigated for potential use as oxidation catalysts. At the same time, mineral formulations such as hopcalite and vanadium pentoxide were tried and discarded. The primary motivation for exploring the use of base metals was lower cost and better availability compared to noble metals. Unfortunately, the base metal formulations were found to be far less active and more sensitive to poisoning than were platinum and palladium. ("Poisoning" is a term used to describe diminished catalyst activity caused by a foreign material that is either deposited on catalyst elements, or reacts with catalyst materials.)

The catalytic activity of base metal oxide catalysts per unit of surface area was found to be lower than that of noble metal catalysts. In other words, for equivalent catalytic performance, more surface area would be required for the base metal oxide catalysts. In addition, the catalytic activity of base metal oxides is severely depressed at temperatures below 600°C by the presence of SO_2 in the exhaust, such as that produced from sulfur-containing compounds in gasoline fuels.

In addition to the efforts to identify acceptable base-metal catalysts, catalyst elements other than platinum, palladium, and rhodium were studied for use in the automotive catalytic converter. Three, in particular, that were examined are iridium, ruthenium, and nickel [6.11].

Iridium has been shown to be catalytically activity in reducing NO to N_2 when the gas stream is oxidizing. Iridium is not used in automotive catalysts today because it is not available in sufficient quantities; it also forms a volatile oxide in the exhaust stream and thus would become depleted in use.

Ruthenium was found to have good catalytic properties in a net reducing gas stream, and was shown to produce little ammonia when reducing NO_x to N_2. Ruthenium was abandoned, however, because its volatile oxide, ruthenium tetroxide, was found to be lost in an oxidizing exhaust stream. Monel™, a Cu/Ni alloy, was tested as a reducing catalyst material. It proved unacceptable because it was found to be chemically attacked by SO_2 in the exhaust gas.

Nickel has been used in some commercial catalysts in combination with platinum, palladium, and copper to promote reduction of NO_x. Nickel forms a stable sulfide in the exhaust and has therefore been used to suppress formation of H_2S in catalytic converters. A problem with nickel is that nickel carbonyl, which might form in the exhaust, has been proven to be carcinogenic. If a catalyst containing nickel were to abrade, or flake off, and be discharged into the atmosphere, it is possible that the nickel-bearing powder could be ingested by humans. This has raised concern among health care providers and environmentalists as to the suitability of nickel as a component of catalytic converters. However, there is controversy among experts about this, because some believe that the nickel carbonyl decomposes at a temperature of approximately 80°C.

The Contributions by Catalyst Companies

Catalysts companies, drawing upon their extensive technical expertise, facilities, and staffs, have made outstanding technical contributions to the development of improved catalyst combinations and washcoat formulations. Companies active during the formative years of catalyst R & D included Engelhard, Johnson Matthey, W.R. Grace, Universal Oil Products, and Air Products. The various catalyst companies invested millions of dollars in manufacturing plants, and developed dedicated bench tests and engine dynamometer tests to support improvements in automotive catalysts. Technical developments contributed by individual catalyst companies, working in conjunction with the automobile companies, have improved both the conversion performance and the durability of catalysts. Thus, catalyst companies have overcome many challenges in providing the automobile industry with the technology necessary for catalytic converters. Technical publications in the open literature describe the results of these efforts. For further reference, an excellent summary of the development of catalyst technology for automobiles is presented in a recent book by Heck and Farrauto [6.12].

Catalysts were developed specifically to perform when exposed to the transient duty cycle imposed by the automobile, and to meet durability requirements. To achieve these aims, new techniques were required to prepare catalysts, involving new stabilizer materials and new washcoat formulations.

Substrates made from both pellets and monoliths have been strengthened to improve crush strength and abrasion resistance. Washcoat formulations and techniques for application have been improved to obtain uniform washcoat distributions with improved stability. Especially important is resistance to sintering, which occurs at high temperatures, and results in surface damage and loss of effective surface area. Sintering is a process whereby the pores of the overheated catalyst collapse, catalyst sites are lost, and the catalyst agglomerates.

The addition of cerium oxide to washcoats has proven to be a major technical contribution, especially for three-way catalysts. Cerium oxide contributes to oxygen storage during lean cycling of the fuel mixture, thereby providing for continued oxidation activity when the fuel mixture cycles rich. In addition cerium oxide improves the thermal stability of the washcoat, and stabilizes the washcoat by inhibiting reactions of noble metals with alumina.

Cerium also enhances the water gas shift reaction and promotes steam reforming. The water gas shift reaction is a reversible, temperature-dependent reaction between CO and water to produce hydrogen and CO_2. It is catalyzed by metal oxides such as ceria. Steam reforming is the reaction between methane or other hydrocarbons and steam, such as are present in automobile exhaust, to produce hydrogen and oxides of carbon. This reaction is also aided by catalysts.

Catalyst companies were concerned about their ability to obtain adequate supplies of noble metals if they would be used extensively in automotive catalytic converters. For this reason, as well as the higher expected costs of noble metals, development efforts by catalysts companies were originally focused on base metals as the preferred catalyst elements. However, base metals were found to be far less active and more sensitive to poisoning than were platinum and palladium. As a result, when oxidizing catalysts were first introduced in 1975, platinum and palladium emerged as the catalyst elements of choice.

Catalyst Aging

Both light-off and conversion performance of catalysts degrade with time in service. The amount of deterioration depends on the formulation of the catalyst, the properties of the substrate, and conditions of the exhaust gas stream supplied to the catalyst. Thermal *sintering* and *poisoning* are the two causes of degradation of catalysts.

When the surface temperature of the catalyst substrate exceeds approximately 850°C, *sintering* deactivates the catalytic activity of many formulations. By the early 1990s, improved technology allowed some catalyst formulations to tolerate 1000°C for extended periods. Sintering is characterized by physical changes to the porous washcoat to the extent that the access of exhaust gases to active catalyst sites is blocked. At the same time, the agglomeration of dispersed metals greatly reduces the number of active sites.

Poisoning describes the presence of elements or compounds in the gas stream that either mask active sites or react with the washcoat materials, catalyst elements, or substrates to decrease catalyst performance. Primary poisoning agents are lead, phosphorous, and sulfur. To counteract performance deterioration resulting from sintering and poisoning, catalytic converters are oversized, a design strategy that ensures adequate performance for the warranted life of the vehicle.

While the automobile manufacturers developed improved fuel metering and improved ignition to control temperatures, the catalyst companies studied improved chemical formulations to stabilize washcoats, making them more resistant to sintering and poisoning. One important finding was that adding barium oxide and lanthanum oxide to an alumina washcoat increased the tolerance of the washcoat to sintering.

Development of Catalytic Converters by the Auto Industry

As 1975 approached, development activities within General Motors AC Spark Plug Division to produce a production catalytic converter accelerated to a hectic pace. As the many samples of catalyst materials began to arrive from suppliers, tests were developed to expedite the sorting and rate the performance of alternative materials. At the same time, development engineers realized that test procedures were needed to compare the conversion performance of catalyst submissions, both fresh and aged, without having to test them on a vehicle. Vehicle testing is not only expensive, but is also subject to a wide range of variables that introduce uncertainties into the test results.

A transient light-off test was one of the first test procedures developed at GM. This was necessary because automobiles are usually started after being parked over night, so all parts are at an ambient temperature, typically 20 to 30°C. Reasoning that exhaust from an operating engine would be necessary even for a screening test, engineers at AC Spark Plug devised a light-off test using exhaust from a spark-ignited engine [6.13]. Tests similar to this, and the others described below, were developed throughout the industry, both at catalyst companies and automobile companies.

To approximate a transient condition from a cold ambient, a test was designed with an exhaust system with two separate pipes. One pipe ducted the exhaust out of the test cell, and the other pipe ducted the exhaust gases to a catalytic converter containing an experimental catalyst and substrate. With an engine operating at a steady-state condition, activation of a quick-acting valve ducted hot gases to the experimental converter. By this method, the test engine could be operated at pre-set conditions so that all catalytic converters were exposed to the same input conditions. This test exposed the catalytic converter to a *pseudo* cold start. This cold start did not duplicate a cold-start transient for any vehicle, but it had the advantage of being repeatable, thereby reducing test uncertainty in comparing alternative catalyst formulations. When the test catalytic converter is fully warm, steady-state efficiencies are also measured.

This test has proven invaluable for rating not only fresh but also aged catalysts. For oxidizing catalysts, this test served both functions, that is, it tested both cold start transient warm-up and final fully warm efficiencies. However, when reducing catalysts for control of NO_X subsequently appeared on the scene, it was found that this test could match either rich or lean operation of the engine, but not both.

Test protocols were also developed to evaluate the durability of conversion performance of the catalyst as well as the durability of the catalyst container itself [6.14]. These test procedures were equally invaluable in solving other durability problems encountered after converters were in production and actually installed in customer's cars.

Durability tests employed an engine-dynamometer facility which arbitrarily set more severe levels for exhaust temperatures and exhaust flow rates than typically found in vehicles in customer use. Results from these tests identified the catalyst formulations with the best physical durability and the best conversion performance after exposure to extremes of vibration and temperature. Durability testing of containers revealed deficiencies in container design or construction. All deficiencies identified were corrected prior to a design being released for production.

In 1975, after more than four years of extensive research and engineering effort to evaluate more than 1000 catalyst formulations from approximately

60 domestic and foreign suppliers, General Motors chose the pelleted converter for production. Test results indicated that the pelleted substrate exhibited good conversion performance and offered somewhat better durability than the more recently developed monolith configuration. In addition, manufacturing processes and equipment had long been in use to produce large quantities of pellets for use by chemical industries, and equipment was in place to coat pellets with catalyst materials.

While GM introduced the pelleted catalytic converter for the 1975 model year, Ford and Chrysler chose the monolithic catalytic converter. The Chrysler Avenger was the first car to meet the 1975 U.S. emission standards, using a rhodium-promoted platinum oxidation catalyst made by Johnson Matthey [6.15]. (See Appendix B for a description of contributors at General Motors, Ford, and Chrysler.)

Sizing the First GM Pelleted Converter

Once a catalyst had been designated for use in an emission control device, the optimum size of the converter unit, i.e., the size of the "container" had to be determined. This was rather complicated, however, because neither GM nor the industry had any previous experience with converters, and there was no known sizing procedure. Before a final decision was made, there was much heated discussion among the engineers, technicians, and GM management. They finally agreed to make the first converter unit the size of the air cleaner for a 1970 vintage engine.

The design included inlet ducts and retaining grids, needed to retain the catalyst pellets, which were approximately 3.2 mm (0.125 in.) in diameter. The container was a circular can approximately 18 inches (45.7 cm) in diameter and four inches (10.16 cm) thick. This size was especially convenient because tooling for stamping sheet metal into this configuration was already being used to make production air cleaners. The dies were simply changed to handle stainless steel instead of carbon steel.

Many converters of this initial configuration and size were built and used for development testing before the final size and shape were decided.

Retaining Grid Dilemma

To keep the pellets in place, it was clear that a retaining device would be have to be installed at both the inlet to and outlet from the pellet bed. Given this assignment, a GM development engineer immediately ordered a sheet of perforated steel to do the job. The sheet contained an array of perforated holes, each hole slightly smaller than 3.2 millimeters (0.125 in.), so that the pellets could not escape from the bed. Unfortunately, however, the holes were still large enough that the pellets could sit in them: When the first catalytic converter was filled with pellets, one pellet sat in every hole, so that the unit became completely plugged and no gas could flow through!

The engineers quickly put this mishap behind them. Several of them developed a perforation grid pattern that worked, and the evolution of the pelleted catalytic converter continued.

GM's 1975 emission control system, shown in Fig. 6-2, included not only a catalytic converter, but also high-energy ignition and an improved early fuel evaporation system. In the bead-bed converter, shown schematically in Fig. 6-3, alumina was pelletized either as spheres or extrudates approximately 3 mm (0.118 in.) in diameter. For several years, American Motors, Avanti, Isuzu and Checker purchased pelleted converters from GM, and some Japanese vehicles were equipped with pelleted converters of their own design.

The first GM catalytic converter, Model 260, released for production in the 1975 model year weighed approximately 10 kg (22.06 lb) and contained 4.26 L (0.150 cu ft) of pellets weighing approximately 2.9 kg (6.40 lb). Impregnated into the surface layers of the pellets was approximately 0.05 troy oz (1.56 g) of a 5:2 mixture of platinum to palladium, essentially duplicating the mixture ratio of these metals as refined from the mine source. The alumina pellets were retained in the converter by perforated metal grids at both the inlet and the outlet faces. An inlet transition duct directed the flow of exhaust gas from the inlet exhaust pipe to the face of the inlet grid; and an exit transition duct directed flow from the exit grid to the connecting downstream exhaust pipe. All of these parts were housed in a stamped, welded metal housing. A smaller model 175 converter was also designed and built.

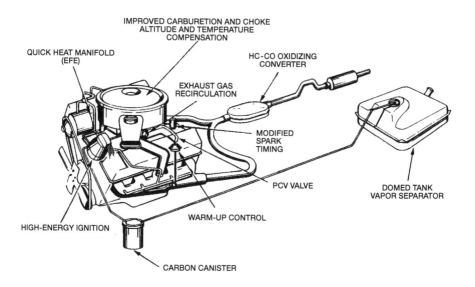

Fig. 6-2 1975 emission control system.

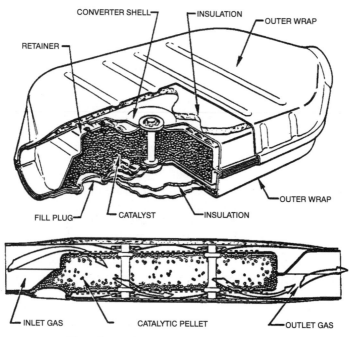

Fig. 6-3 Pelletized catalytic converter.

Manufacturing Catalytic Converters at General Motors

General Motors was faced with the awesome prospect of mounting a pelleted catalytic converter on each of the approximately five million automobiles it would build for the 1975 model year, not to mention the increasing numbers it expected to sell in following years. These catalytic converters would be exposed to a myriad of environmental conditions, used on a variety of vehicle sizes, and operated by a wide cross section of consumers. To accomplish this complex task, a monumental production effort was initiated. Contracts were negotiated with several major manufacturing companies, including The Catalyst Company (a joint venture between American Cyanamid and Japan Catalytic Chemical Inc.), W. R. Grace & Co., Air Products, and Engelhard, to supply *tons* of pellet-supported noble metal catalysts.

To fulfill GM's requirement for up to 300,000 troy ounces of noble metals per year, an agreement was consummated with Impala Platinum, Ltd. of South Africa to supply 300,000 troy ounces of platinum and 120,000 troy ounces of palladium annually, beginning in 1975. To supply these quantities of noble metals, 10,000 miners were added to the work force to mine ore from the Merensky Reef area of South Africa. For each troy ounce of platinum and palladium mixture ultimately deposited in a catalytic converter, approximately twelve tons of rock had to be mined.

General Motors' entire passenger and light truck vehicle line-up was equipped with pelleted catalytic converters manufactured by AC Milwaukee Operations, located in Oak Creek, Wisconsin. Francis J. Buckley, Jr. was named manager of the AC Milwaukee plant on March 1, 1973. The company's objectives were to renovate existing buildings and erect new structures to produce a target quantity of 30,000 catalytic converters per *day*. Each day, a long freight train or trucks of equivalent capacity hauled raw materials into the plant while another carried out the finished product. The operation required 300 tons of steel per *day*, and would ultimately occupy 889,000 square fleet of floor space. This included a 24-press bay with 16 huge stamping presses with capacities up to 1800 tons, which were used to form converter housings and retainers from a special chromium stainless steel.

Through 1995, approximately 130 million catalytic converters were manufactured at the Oak Creek Plant; during this time the design slowly migrated from the pellet to the monolith type. Production continues into the 21st century, although during the 1990s production declined somewhat as other automotive supplier companies built production facilities to compete for the business.

The Ceramic Monolith Catalytic Converter

While pellet makers were developing heat-resistant substrates and washcoats, the Minnesota Mining and Manufacturing Company (3M) and Ford Motor Company were experimenting with corrugated designs in which thin papers were dipped in ceramic slurries, corrugated, then fired to make a ceramic "honeycomb." These development efforts were the first steps toward producing a monolith catalytic converter, which, compared to the pelleted configuration, offered a packaging advantage and lower pressure drop.

The monolith catalytic converter, shown in Fig. 6-4, appears on the outside to be very similar to the bead bed converter, but inside it is totally different. The honeycomb structure of a monolith catalytic converter substrate, which contains many small parallel passages, is formed by an extrusion process.

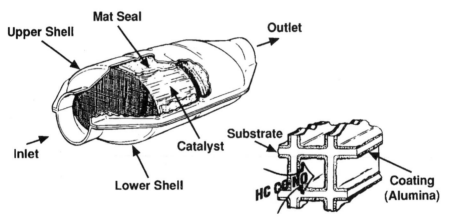

Fig. 6-4 Ceramic monolith catalytic converter.

The flow passages of a monolith catalytic converter are usually square or triangular in cross section, although other cross sections have been extruded. To provide a support for the noble metals in a monolith converter, the very low porosity cordierite walls are coated with 15 to 25 microns of porous alumina "washcoat." Next, catalyst materials are impregnated into the washcoats via the small monolith channels. The term "monolith" is used to describe such a substrate because each individual passage extends all the way from the inlet of the substrate to the outlet at the other end.

Making a monolith converter substrate requires manufacturing processes that are totally different from those used to make pellets. Processing and materials developments by the ceramic industry, for example by Corning Glass Works and Nippon Gaishi Kabushiki (NGK), have resulted in a low-thermal-expansion cordierite material that can be extruded into uniform passages. Continuing development has led to the production of monoliths with excellent dimensional control as well as prototype monolithic substrates with higher cell densities and thinner walls.

For the same pressure drop, the monolith geometry needs less frontal area than a pelleted geometry and requires less volume for the same conversion efficiency. Overall, in comparison to the pelleted configuration, the monolith geometry offers the advantages of smaller volume, less mass, and greater ease of packaging. As monolith technology improved, the product offering at GM slowly migrated to ceramic monoliths.

Recognizing the packaging and sizing advantages, GM introduced monoliths in limited production on GM automobiles in 1977. By 1993, approximately 80% of GM light duty vehicles were equipped with monolithic catalytic converters. The pelleted geometry proved to be more robust for the more severe operating conditions required for trucks. GM continued to use pelleted converters on trucks into the 1990s.

Catalyst Processes

As shown in Fig. 6-5, an oxidizing catalyst promotes the oxidation of HC and CO at threshold gas temperatures between 250 and 300°C, much lower than the 720 to 750°C minimum temperatures required for gas-phase reactions. When a fresh catalyst is fully warm, conversion efficiency for CO approaches 99% and that for HC approaches 96%, depending on the hydrocarbon mix in the exhaust. Fully warm conversion efficiency for CO is higher than that for HC because some of the hydrocarbon species, such as methane, are virtually unreactive in the presence of a catalyst.

The characteristics of the washcoat, in particular washcoat composition, structure, and deposition technique, are as important as the choice and amount of catalyst elements. Figure 6-6 shows a schematic of the microscopic

Metal Monolith Development

The metal monolith catalytic converter, shown in Fig. S6-1, is a recent development that uses a substrate manufactured from very thin, 0.05 mm (0.002 in.) thick, metal foil made of a special high-temperature steel alloy containing chromium and aluminum. A whisker oxide on the metal surface securely bonds to a ceramic gel to retain the alumina washcoat.

The use of a very thin metal foil offers the potential for lower pressure drop when compared to that for a ceramic. In addition, the metal foil substrate can be assembled in many ways and sized into a variety of shapes more easily than a ceramic substrate, which must be extruded.

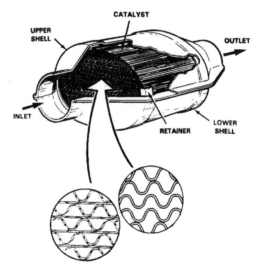

Fig. S6-1 Metal monolith catalytic converter.

Figure S6-1 illustrates two forms of metal monolith substrate construction: a "plate-fin" and a "herringbone." In the plate-fin construction, straight corrugated sheets are separated by flat plates. In the herringbone construction, angled corrugations maintain the separation of the corrugated sheets and no separators are needed. The plate-fin construction is easier to fabricate than the herringbone; however, the herringbone construction requires less foil to provide mass-transfer performance equivalent to that of the plate-fin. This advantage is attributed to the absence of "corners" in the flow passage, i.e., essentially all herringbone surface area is directly exposed to flow of exhaust gas.

GM pioneered the herringbone metal monolith catalytic converter. In the late 1980s, the AC Rochester Division of General Motors developed a metal-foil substrate for catalytic converters. GM, within the company, further developed the processing for this design, including foil fabrication, substrate retention, washcoat application, and catalyst impregnation [S6.9]. AC Rochester delivered production metal monoliths to Opel beginning in 1989. Although the product was successful in terms of performance and durability, after several years the metal monolith program was halted, the primary reason being economic: the cost of the metal substrate just wasn't competitive with the cost of an equivalent ceramic monolith.

Several companies in addition to AC Rochester have developed metal mono-liths, including, Camet (a subsidiary of W. R. Grace), Behr, Kemira, and Emitec [S6.10, S6.11, S6.12, and S6.13]. These companies have had varying degrees of success in the marketing of metal monoliths. Most of the early applications were on high-performance vehicles where space constraints favored the mono-lith configuration. Emitec has emerged as one of the most active suppliers of metal monoliths, supplying catalytic converters to Audi, BMW, Fiat, Jaguar, Rolls Royce, Scania, and Volkswagon in Europe. In the United States, Emitec began to supply small light-off converters installed on Chrysler L and H vehicles beginning in 1992; and metal monoliths supplied by Emitec were included to control warm-up emissions on GMT800 trucks released in 1998.

References

S6.9 Vaneman, G.L., "Performance Comparison of Automotive Catalytic Con-verters: Metal vs. Ceramic Substrates," XXIII FISITA Congress, Vol. I, pp. 865–870, Technical Paper 905115, May 1990.

S6.10 Nonnenmann, M., "Metal Supports for Exhaust Gas Catalysts," SAE Paper No. 850131, Society of Automotive Engineers, Warrendale, Pa.,1985.

S6.11 Pelters, S., Kaiser, F.W., and Maus, W., "The Development and Application of a Metal Supported Catalyst for Porsche's 911 Carrera 4," SAE Paper No. 890488, Society of Automotive Engineers, Warrendale, Pa., 1989.

S6.12 Maattanen, M, "Mechanical Strength of a Metallic Catalytic Converter Made of Precoated Foil," SAE Paper No. 900505, Society of Automotive Engineers, Warrendale, Pa., 1990.

S6.13 Bruck, R., Diringer, J., Martin, U., and Maus, W., "Flow Improved Efficiency by New Cell Structures in Metallic Substrates," SAE Paper No. 950788, Society of Automotive Engineers, Warrendale, Pa., 1995.

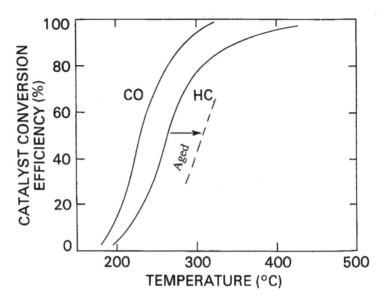

Fig. 6-5 Light off for a typical catalyst, including the effect of aging.

structure of the washcoat, which reveals catalyst pores, macropores, and micropores [6.16]. Very fine noble metals particles "dispersed" or deposited onto the walls of these pores function as the catalyst elements. As mentioned previously, washcoats also contain *promoters* and *stabilizers*, usually oxides of metals such as ceria, nickel, iron, and copper.

Molecules of HC, CO, and NO_X are transported from the free stream to the exterior "surface" of the flow passage by mass transport, shown schematically in Fig. 6-7. The mass transport process is well understood as a result of research studies conducted at GM Research and other research laboratories during the time period from 1970 to 1990 to correlate the mass transport coefficient [6.17]. Unfortunately, mass transport controls the conversion efficiency only when a fresh catalyst is fully warm. Under most operating conditions, conversion performance of an automotive catalyst depends on a combination of mass transport and "catalyst kinetics."

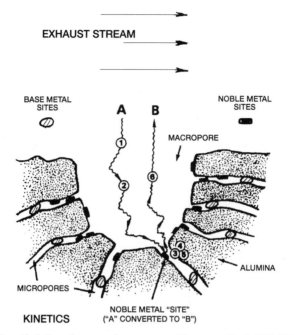

Fig. 6-6 Catalyst surface pore. (Source: Ref. [6.18].)

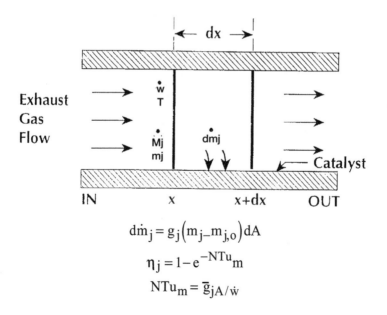

$$d\dot{m}_j = g_j\left(m_j - m_{j,o}\right)dA$$

$$\eta_j = 1 - e^{-NTu_m}$$

$$NTu_m = \bar{g}_j A / \dot{w}$$

Fig. 6-7 Mass transfer process.

The several fundamental kinetic processes shown in Fig. 6-6 by numbers 1 through 6 are:

1. A molecule of "A" must diffuse to the surface.
2. It diffuses within the macropore.
3. "A" is adsorbed at an active site.
4. Reaction transforms "A" into "B."
5. Molecule "B" is desorbed.
6. Molecule "B" diffuses back through the macropore to the gas stream.

Stanford Kinetics Study

AC Rochester, working with the Physical Chemistry Department at GM Research, joined with the National Science Foundation (NSF) to support a four-year contract research program at Stanford University, beginning in 1990 [S6-14]. The aim of this research program was to study the diffusion and kinetic processes occurring within monolith passages. For this study, a scaled experiment was designed using two parallel flat-plates, spaced 2.0 cm (x in.) apart, to simulate the geometry of a monolith passage. Using a special heating and cooling scheme, both walls were maintained at a constant temperature. This eliminated the influence of temperature changes, and the stable environment permitted more accurate determination of kinetics of species of interest.

In this study, the spacing of the plates allowed access to the flow stream, and permitted species profiles in the boundary layer of the gas stream to be sampled. From these boundary layer profiles, the following kinetic rate constants were measured:

$$R_{CO} = \frac{K[CO][O_2]}{\left(1 + K_{CO}[CO] + K_{O_2}[O_2]\right)}$$

$$K = k_i e^{-E/RT} \qquad K_{CO} = k_{CO} e^{-E_{CO}/RT} \qquad K_{O_2} = k_{O_2} e^{-E_{O_2}/RT}$$

The Stanford studies identified the presence of a "kinetic-controlled" zone in the upstream portion of a fully warmed monolith passage, followed by a "diffusion-controlled" zone in the remaining downstream portion, as shown in

Detailed theories, mathematical models, and experimental confirmations exist for each of these processes. Although complex, kinetic theory can be used for either oxidizing or reducing reactions. A schematic of a simplified kinetics process, which can be used for empirical sizing, is depicted in Fig. 6-8.

In Fig. 6-9, an Arrhenius plot [6.18] shows a correlation for the kinetic rate coefficient for CO oxidation. This demonstrates the strong dependence of the rate coefficient on both temperature level and catalyst age.

Today's automotive catalyst is a three-way catalyst, combining simultaneous oxidizing and reducing reactions into one product. Because of this three-way catalysis, plus multiple reactions and the A/F dynamics in an automobile

Fig. S6-2. This finding suggested the possibility of modeling fully warmed catalysis by treating a monolith passage as two different zones. The upstream zone would be rate-limited by catalyst kinetics, and the downstream zone rate-limited by mass-transport processes. The parameters that control the fraction of monolith that is rate limited have not been defined for either oxidizing or three-way operation.

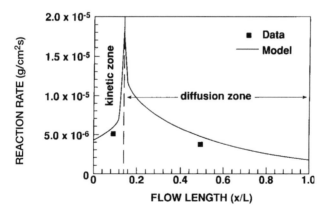

Fig. S6-2. Kinetics and diffusion. (Source: Ref. [S6.14]).

Reference

S6.14 Boehman, A., Niksa, S., Moffat, R, "Catalytic Oxidation of Carbon Monoxide in a Large Scale Planar Isothermal Passage," SAE Report No. 922332, Society of Automotive Engineers, Warrendale, Pa., 1992.

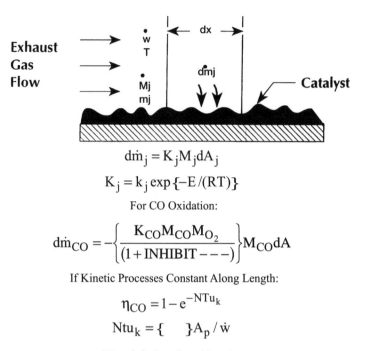

$$dm_j = K_j M_j dA_j$$

$$K_j = k_j \exp\{-E/(RT)\}$$

For CO Oxidation:

$$dm_{CO} = -\left\{\frac{K_{CO}M_{CO}M_{O_2}}{(1+INHIBIT---)}\right\}M_{CO}dA$$

If Kinetic Processes Constant Along Length:

$$\eta_{CO} = 1 - e^{-NTu_k}$$

$$Ntu_k = \{\quad\}A_p/\dot{w}$$

Fig. 6-8 Catalyst kinetics.

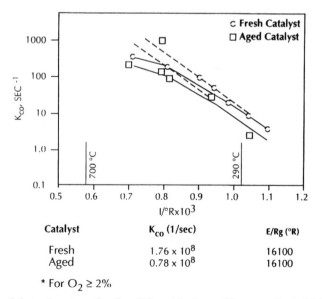

Catalyst	K_{co} (1/sec)	E/Rg (°R)
Fresh	1.76×10^8	16100
Aged	0.78×10^8	16100

* For $O_2 \geq 2\%$

Fig. 6-9 Arrhenius plot for CO oxidation. (Source: Ref. [6.20].)

system, accurate modeling of such a system based on theoretical analysis is not possible. However, the conversion performance of a catalytic converter can be measured as a function of volume by cutting a monolith into segments and measuring the performance of the different portions. The results of a study of the performance of segments of monoliths are presented in Figure 6-10. From this figure, it can be seen that most of the conversion is completed in the front sections of a catalytic converter, even after aging for 30,000 miles.

1975 Catalytic Converter Emission Control System

In 1975, in order to satisfy U.S. regulations that established target emission levels of 1.5 g/mi for HC and 15 g/mi for CO, catalytic converters were introduced on essentially all passenger cars sold in the U.S. Prior to the introduction of the catalytic converter, emissions control systems had relied on retarded spark timing and injection of additional air into the exhaust to promote the oxidation of HC and CO. Unfortunately, the use of retarding

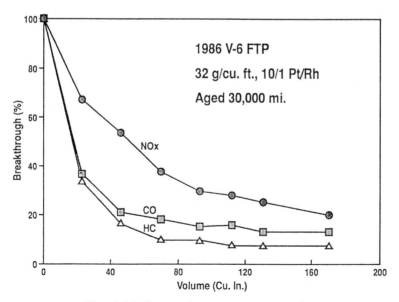

Fig. 6-10 Conversion vs. converter volume.

spark timing to lower emissions resulted in a substantial penalty in fuel economy. Depending on the specific vehicle and emission control required, this penalty amounted to as much as 10 to 12%.

Installing a catalytic converter to promote the oxidation of HC and CO allowed the spark timing to be advanced to a setting approaching minimum spark advance for best brake torque (MBT). This spark setting improved not only fuel economy, but driveability as well. Due in large part to the introduc-

Fuel Economy Standards

In 1975, the United States Congress passed the Energy Policy and Conservation Act (EPCA), establishing the Corporate Fuel Economy Standards (CAFE), and marking the first time the government had ever assumed a role in regulating the fuel economy of passenger cars. Beginning with model year 1978 vehicles, EPCA required manufacturers to meet a production-weighted average fuel economy of 18 miles per gallon for domestic passenger cars. An ultimate target of 27.5 miles per gallon was stipulated for model year 1984, which would   essentially double the average fuel economy measured in 1974 (S6.15). The auto companies faced fines if their vehicle fleets failed to meet the established standards.

Recognizing that projecting fuel economy standards for 1985 from the vantage point of 1975 might not be wholly feasible, Congress authorized the Department of Transportation (DOT) to set the standards for the interim years of 1981–84. The target standards for passenger cars prescribed by DOT for these years subsequently were established at 27 miles per gallon for 1984 and 27.5 miles per gallon for 1985. The auto manufacturers responded to the mandated fuel economy requirements by installing more efficient engines, transmissions, and accessories, and by reducing vehicle mass.

Engine efficiencies were improved by means of more efficient inlet manifolds to improve breathing, more accurate fuel delivery with the use of fuel injection systems, and improved controls such as fuel shut-off during engine decelerations. Electric fans replaced belt-driven fans, reducing wasted power by several horsepower; and coolant pumps were redesigned to be more efficient.

Friction losses were reduced in the valve train, transmission, drive train, and brakes; brake drag, especially, was reduced. Other transmission improvements

tion of the catalytic converter, the average sales-weighted fuel economy for the GM vehicle fleet improved from 12.0 to 15.4 miles per gallon in 1975, as shown in Fig. 6-11. However, some of this improvement in fuel economy can be attributed to other improvements such as tires with reduced rolling resistance.

The vacuum hose schematic for a 1980 GM 4-cylinder engine (Fig. 6-12) is typical of the sophisticated emission control systems on vehicles equipped

---

included "built-in" overdrive gear ratios permitting engines to operate more efficiently at road load conditions; and new components to improve driveline efficiency. In addition to all of these internal improvements, tire manufacturers developed a new generation of tires with reduced rolling friction.

GM met the passenger car CAFE standards until model year 1983, when the price of gasoline declined somewhat, and customer demand for larger cars increased. Although GM failed to meet the standards for the period 1983–85, credits from other years were used to offset the penalty and GM was not required to pay fines.

The U.S. auto industry petitioned DOT to change the CAFE standard to 26 miles per gallon beginning in 1986. The logic behind this was that retaining the higher standard would allow certain foreign manufacturers, who held substantial CAFE credits due to their large fraction of small car sales, to use credits to export their larger vehicles, and these imports to the U.S. would reduce domestic employment and cause economic disruption. Reacting to concern about potential damage to the U.S. economy, DOT established fuel economy standards of 26.0 from 1984 through 1989, and the original target of 27.5 was postponed to model year 1990.

Domestic light trucks also had to meet legislated fuel economy standards. For two-wheel and four-wheel drive trucks, the initial standard of 20 miles per gallon was set for the 1990 model year, and was subsequently increased to 20.5 beginning in 1998.

Reference

S6.15 Case, D.E., "The CAFE Standards: Ten Year Later," SAE Paper No. 852373, Society of Automotive Engineers, Warrendale, Pa., 1985.

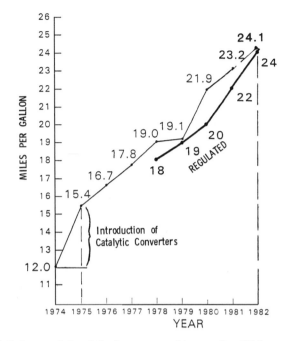

Fig. 6-11 Sales-weighted fuel economy history for GM passenger cars.

with catalytic converters [6.19]. As shown in this figure, the carburetor is the focal point, and other devices are arranged as "satellite" components.

1. Distributor spark-vacuum delay (DS-VDV).
2. Distributor spark-vacuum modulator valve (DS-VMV).
3. Exhaust gas recirculation-thermal vacuum switch (EGR-TVS).
4. Check valve in exhaust for secondary air (PULSAIR).
5. Positive crankcase ventilation (PCV).
6. Deceleration valve.
7. Charcoal canister for control of fuel evaporation.
8. Exhaust gas recirculation (EGR) valve.

A side view of the carburetor body in the upper left of Fig. 6-12 shows several ports in the carburetor venturi provided for access to different levels of manifold vacuum, as well as the bowl vent, which vents vapor to the charcoal canister. "Spark port" designates a source for manifold vacuum

118

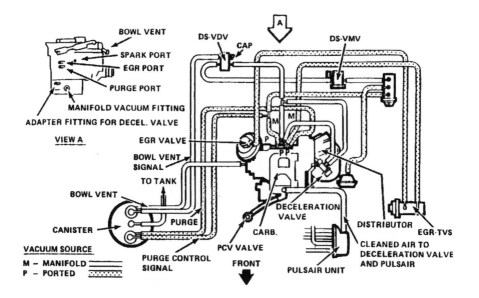

Fig. 6-12 Vacuum hose schematic, 1980 4-cylinder engine.

located partway up the side wall of the venturi; at this location the port is exposed to manifold vacuum only after the throttle plate rotates past the port. Thus, when the throttle is closed or slightly open, there is no vacuum signal available to modify spark advance.

Catalysts for Diesel Engines

Catalytic converter systems to control emissions from gasoline engines were in use several years before converters were also considered for control of exhaust emissions from diesel engines. Exhaust emissions from diesel engines are more complex than those from gasoline engines, and include solids, liquids, and gases. The solids are carbon particles, or particulates, also known as soot. The liquids are unburned diesel fuel, and lubricating oils, which are identified as soluble organic fractions (SOF). Gaseous pollutants include unburned hydrocarbons, carbon monoxide, oxides of nitrogen, and sulfur dioxide. Some of the sulfur in the fuel is oxidized to sulfate, which can react with water vapor to form sulfuric acid.

119

The appeal of diesel engines is based on their high fuel efficiency and long life as compared with gasoline spark-ignited engines. Beginning in the mid-1980s, manufacturers of diesel engines developed improved fuel injectors, combustion chambers, and timing control. These modifications were necessary in order to meet a particulate standard of 0.25 grams per brake horsepower hour, as required in 1990 by EPA. This particulate standard was measured using the EPA heavy-duty test cycle for diesel engines. Similar levels of emission control for particulates were required in Japan and Europe. A new particulate standard of 0.1 grams per brake horse-power hour was required in 1994. In order to meet this new standard, the diesel industry realized that catalytic converters were the only feasible solution for most diesel installations.

Very lean operation of diesel engines produces little HC and CO, so the challenge for diesel engine designers is to reduce both NO_X and particulates. Engine modifications and reductions in the sulfur content of fuel have decreased particulate emissions substantially. This has increased soluble organic fraction (SOF) content, but, fortunately, SOF emissions can be oxidized with a catalytic converter. A catalytic converter for diesel exhaust must meet two major criteria: (1) It must control SOF at low exhaust temperatures, and (2) It must be selective so as not to oxidize SO_2 to SO_3, which is a precursor in the formation of sulfuric acid.

To control SOF at low temperatures, the catalyst and washcoat surface layer adsorbs the SOF. Then, when the exhaust temperature increases as a result of engine operation, the adsorbed SOF can be oxidized by a catalyst. To minimize formation of SO_3, oxides of titanium, silicon, and zirconium were substituted for the aluminum oxide washcoats used in gasoline catalytic converters. This substitution was necessary because alumina washcoats react with SO_2 to deactivate the catalysts. Combinations of Pt and Pd catalyst elements with improved washcoats and washcoat additives have resulted in catalytic SOF conversion efficiencies of more than 50%.

Although EGR was somewhat effective in controlling NO_X emissions, stringent emissions targets required significant additional lowering of NO_X. To further reduce NO_X, in recent years much effort has been spent on developing a lean-burn NO_X catalyst. Chapter 7 provides additional information these developments.

Temperature Effects

The primary function of a catalytic converter is to lower the temperature at which HC, CO, and NO_X gases react at the catalyst sites. A major problem is that maximum temperatures can degrade catalyst performance. Effects of operating temperature on catalysts are summarized in the "thermometer" chart presented in Fig. 6-13.

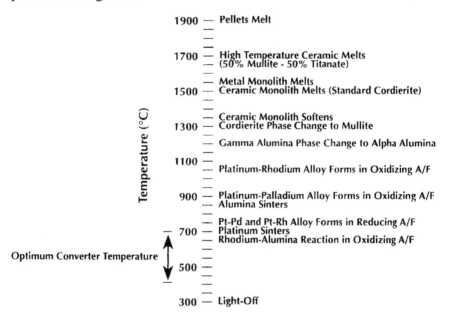

Fig. 6-13 Effect of operating temperatures on catalysts.

Catalyst activity requires a light-off temperature of approximately 300°C. The preferred "normal" operating temperature ranges between 400 and 700°C. When temperatures exceed 750°C, noble metals chemically react both with each other and with alumina in the washcoat, the outcome being a deterioration in catalyst performance. And when temperature increases to 1200°C, gamma alumina changes to alpha alumina with attendant washcoat shrinkage, loss of micropores, and a tenfold reduction in catalyst surface area. Above 1300°C, the substrate itself suffers thermal damage. To avoid some of these problems, new catalyst formulations were developed. By the mid-1990s, palladium catalyst formulations were available which could withstand temperatures up to 1000°C.

121

The oxidation reactions for HC and CO are both exothermic, thus engine misfiring or rich engine operation can generate high temperatures in a catalytic converter. A warmed-up catalyst will completely oxidize in milliseconds over 90% of the unburned HC and CO, generating a certain adiabatic flame temperature in the gas. This flame temperature can be estimated from the concentrations of CO and unburned HC in the exhaust, as graphed in Fig. 6-14. The adiabatic flame temperature is a good measure of substrate temperature. The substrate offers almost negligible thermal storage for the released energy, and there is insufficient time for heat transfer away from the inside of a converter. Furthermore, when the temperature of the gas mixture exceeds 750°C, gas-phase reactions begin to occur which accelerate the oxidation combustion process.

As shown in Fig. 6-14, rich operation of an engine with 1% CO and sufficient oxygen present can add 111°C to the operating temperature of a catalyst. And, on top of this, there is a temperature increase resulting from the oxidation of HC. If the concentration of HC is 1000 ppm as C_6 (hexane), an additional 111°C will be added. During rich operation of an engine, additional oxygen required for oxidation is usually supplied by an air pump. With 25% engine misfire, a typical concentration of 5000 ppm hexane will reach the converter along with the necessary oxygen. When this gas mixture

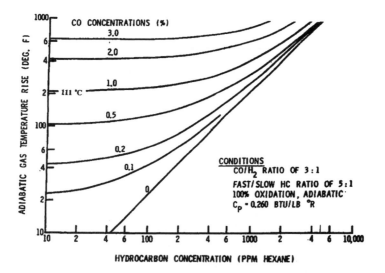

Fig. 6-14 Theoretical adiabatic gas temperature rise.

reacts, the local operating temperature will increase by approximately 555°C. With 25% misfiring, the catalytic function of the converter will suffer; with more than 33% misfiring, localized melting of the substrate can be expected.

Fig. 6-15 shows a schematic of the microscopic structure of a catalyst that has undergone thermal damage by "sintering" and "agglomeration." Sintering eliminates micropores and the associated catalyst surface area. Agglomeration of the dispersed noble metals at high temperatures reduces the number of active catalyst sites. Conversion efficiencies from a sweep test for a three-way catalyst are presented in Fig. 6-16 [6.20]. In this figure, the performance of a three-way catalyst aged at 700°C is compared with that of the same catalyst aged at 870°C. For this particular catalyst with nickel and cerium stabilizers, HC conversion suffered very little when the aging temperature was increased. However, CO and NO_X conversion efficiency suffered drastically when aging temperature increased from 700°C to 870°C.

Deterioration of catalyst performance has been measured for catalysts aged using a rapid aging dynamometer test procedure. After aging, the performance sweep test is used to measure conversion efficiencies. Fig. 6-17 shows the impact of temperature on converter performance, represented by HC

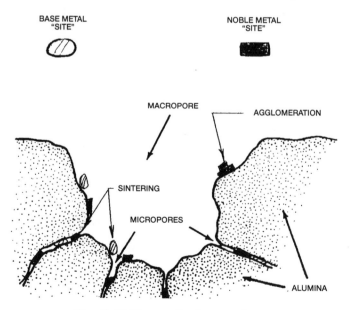

Fig. 6-15 Thermally damaged pore.

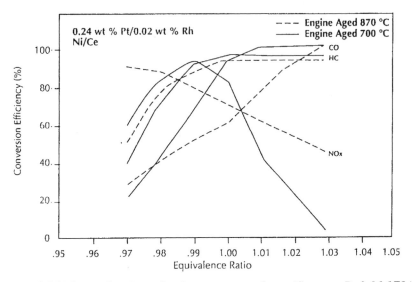

Fig. 6-16 Thermal aging of a three-way catalyst. (Source: Ref. [6.17].)

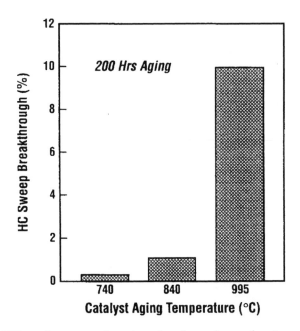

Fig. 6-17 HC performance deterioration from thermal aging. In this figure, "breakthrough" refers to the fraction of incoming constituent remaining after passing through the catalytic converter.

conversion efficiency, as measured by the sweep test at the stoichiometric A/F. This correlation shows that when a catalytic converter is subjected to operating conditions that result in catalyst temperatures exceeding 800°C, rapid deterioration of performance results.

Limitations on Engine Operation

As discussed previously, engine control can strongly impact converter deterioration. Additives in fuels and oils also have the potential to cause catalyst deterioration. In addition, very low levels of lead can poison the catalyst, and, as explained above, high temperatures cause thermal degradation. Table 6-1 summarizes the parameters related to fuels, oils, and engine operation that must be controlled to minimize catalyst performance deterioration.

Table 6-1. Parameter Limits to Minimize Catalyst Deterioration

Fuel lead limit: 0.002 g/L

Engine oil consumption: <0.12 L per 1000 km

A/F perturbations: < ±0.5 A/F at frequencies faster than 1 Hz

Allowable Maximum Catalyst Solid Temperatures:

Percent of Operation	Temperature (°C)
>90	400-700
<10	700
<3	750
<1	800
0	850

Temperature guidelines for converter components are summarized in Fig. 6-18, categorized according to substrate, washcoat/catalyst, and support materials. The guidelines summarized in this figure were appropriate until the early 1990s; by the late 1990s, the allowable temperatures of components had increased as a result of continued development of materials. By using more expensive, premium materials for substrate, washcoat, and catalyst, operating temperatures can now be extended to levels higher than 1000°C.

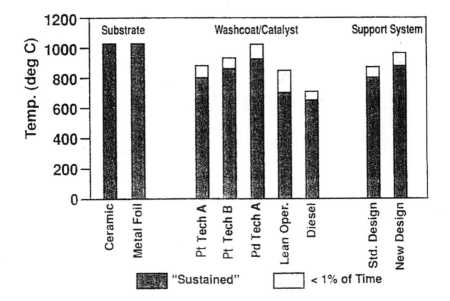

Fig. 6-18 Temperature guidelines for converter components.

Three-Way Catalytic Converter Era

Beginning in 1981, allowable levels for HC, CO, and NO_X emissions were all lowered substantially by new standards set by the EPA. Also included in these new standards was a particulate standard of 0.2 g/mi, to take effect in1986, and a formaldehyde standard of 0.015 g/mi, to take effect in 1993.

A catalyst approach to lowering all of the pollutants required oxidizing HC and CO while simultaneously reducing NO_X. The primary chemical reactions involved are shown below. These equations are approximate because there are many hydrocarbons with ratios of H/C other than one and NO_X represents a variable amount of oxygen.

Three-way catalysts:

$$2HC + CO + 2NO_X + O_2 \rightarrow H_2O + 3CO_2 + N_2$$

Chemical oxidation (HC and CO):

$$2CO + O_2 \rightarrow 2CO_2$$

$$4HC + 5O_2 \rightarrow 2H_2O + 4CO_2$$

Chemical reduction (NO_X):

$$4NO + 4CO \rightarrow 2N_2 + 4CO_2$$

$$4HC + 10NO \rightarrow 5N_2 + 2H_2O + 4CO_2$$

$$CO + H_2O \rightarrow CO_2 + H_2$$

$$2NO + 5H_2 \rightarrow 2NH_3 + 2H_2O$$

$$2HC + 3H_2O \rightarrow CO + CO_2 + 4H_2$$

To achieve the level of emissions control required by these new standards, auto manufacturers and catalyst companies reviewed all emissions control technologies previously investigated. It was immediately evident that the current level of technology was inadequate and that a new approach was needed. The approach selected was a three-way catalyst (TWC) system. The three-way catalyst is so named because one catalyst formulation simultaneously promotes oxidizing and reducing reactions to control all three exhaust pollutants. Components of a 1981 emission control system are shown in Fig. 6-19:

1. Three-way catalytic converter (TWC)
2. Oxygen sensor
3. Electronic control module (ECM)
4. Electronic ignition
5. Closed-loop carburetor
6. Early fuel evaporation (EFE)
7. Exhaust gas recirculation (EGR)

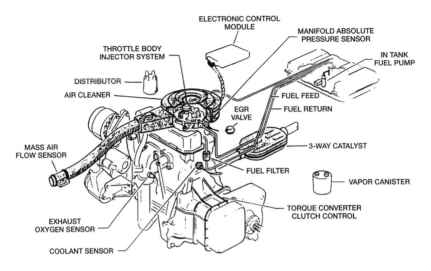

Fig. 6-19 Closed loop emission control system.

Three-Way Catalyst Transient Testing

Cycling tests, which cycle and modulate A/F, were developed by the auto industry and the catalyst companies to compare fully warm performance of three-way catalysts. The test used extensively at General Motors, developed by its AC Spark Plug Division, used an engine dynamometer to expose the catalyst to a mixture of gases produced by an operating engine [6.23, 6.24]. The A/F for the engine was deliberately cycled at a frequency of 1 Hz and an amplitude of approximately ±0.5. This fixed cycling of A/F continued while the net mean A/F was slowly modulated from a lean to rich value of ±0.5 about the stoichiometric A/F. By this method, the dynamometer-controlled test subjected the three-way converter to a combination of cycled A/F along with a variation in mean A/F, including both rich and lean operation. This test has become known as the "sweep" performance test. However, as was the case with the transient light-off test, the sweep test does not duplicate the cycling transient conditions for any vehicle. Nonetheless, results from this sweep test were used to compare the conversion efficiency, both fresh and aged, of alternative catalytic converters and catalysts.

To rate alternative catalytic converters for emissions control on a vehicle, the conversion performance for both fresh and aged catalysts could be evaluated by an engine dynamometer test, which simulated the official federal test procedure (FTP) for measuring emissions. The FTP required that a complete vehicle be tested while being driven on a chassis dynamometer. The dynamometer FTP simulation matched all features of the vehicle FTP test except the exact cold start, which varies from vehicle to vehicle. The engine-dynamometer test eliminated much of the variability of a vehicle test; it also was a much better test for comparing alternative catalytic converters. Further, it could be completed much faster and at much lower cost than a vehicle test.

Continued development of durability testing by catalyst and automobile companies has resulted in "rapid aging test" protocols to age catalysts quickly. At General Motors, the rapid aging test used a three-step dynamometer protocol. By means of this test, a catalyst was aged in 800 hours to simulate 50,000 miles of vehicle durability.

Closed-Loop Emission Control

By adding rhodium to the platinum and palladium catalyst mixture, NO_X can be reduced while both HC and CO are simultaneously oxidized. Concurrent reducing and oxidizing is possible if the air-fuel ratio (A/F) is maintained within a narrow band of ±0.2 A/F about the stoichiometric A/F, as shown in Fig. 6-20.

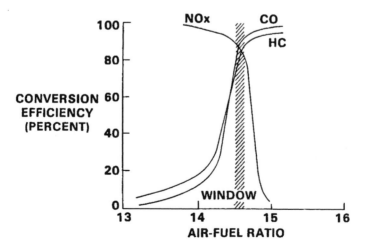

Fig. 6-20 Conversion efficiencies for a typical three-way catalyst.

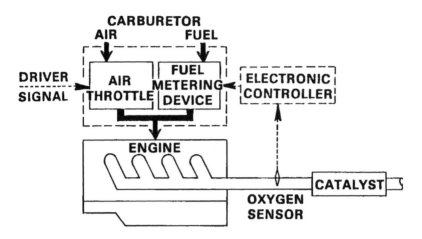

Fig. 6-21 Schematic diagram of a closed-loop control system.

With a "conventional" carburetor, this narrow band of A/F control was determined impossible to achieve and maintain for any significant life. Computer control of fuel metering using a feedback control system, as depicted in the block diagram in Fig. 6-21, was ultimately used to achieve this precise control of A/F.

To control fuel input, an electronic control module (ECM) responds to a signal from the exhaust oxygen sensor. The fuel-metering component provides the correct amount of fuel needed by the engine to maintain the correct A/F based on the amount of air supplied to the engine. To control engine power, the vehicle operator modulates the throttle to control the amount of air supplied to the engine.

An oxygen sensor mounted in the exhaust stream provides the feedback signal to the closed-loop system that controls fuel metering to the engine. This platinum-coated zirconium oxide sensor, shown in Fig. 6-22, acts as a galvanic cell which compares the oxygen level in ambient air with that in the exhaust stream. Output from this sensor is a millivolt signal, which switches at stoichiometric A/F and produces approximately 1000 mV when the exhaust is rich and 10 mV when the exhaust is lean.

Due to finite gas transport times and the on/off signal from the oxygen sensor, the tailpipe A/F is continuously cycling from rich-to-lean-to-rich. This

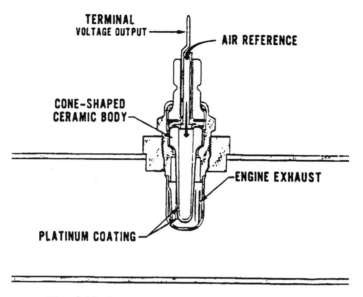

Fig. 6-22. Oxygen sensor in the exhaust pipe.

continuous cycling makes possible efficient three-way catalysts because oxygen storage by the catalyst allows oxidation to continue even when the average A/F of the mixture A/F is rich. The addition of oxides of cerium to the catalyst has proven a very effective means of promoting "oxygen storage" for this three-way catalyst cycling operation.

Many combinations of cycling frequency and amplitude of A/F were investigated by auto and catalyst companies. A/F modulation amplitudes and frequencies were varied between ±2 and ±10 A/F and 0.5 and 5 Hz. The results from these studies indicated that the optimum conversion efficiencies for all three constituents occurred a frequency of 1 to 2 Hz and amplitude of approximately ±0.5 A/F [6.23].

The electronic control module (ECM) processes the oxygen sensor signal and adjusts fuel supplied to the air stream. Fuel rate is adjusted to maintain a mixture of air and fuel at the stoichiometric value ±0.2 A/F. A modified carburetor with oscillating metering rods was the first radically different fuel-metering component to be used by GM in the feedback control system. This electromechanical carburetor was used only a few years before it was replaced by the throttle body injection (TBI) carburetor.

The TBI carburetor, shown in Fig. 6-23, is a hybrid of a carburetor and a fuel injector. The TBI system retains the carburetor's venturi and throttle to control air flow and therefore engine power; however, fuel is injected into the air stream by an electrically operated injector mounted above the throttle. The fuel injector is driven by a solenoid such that it oscillates between a lean and a rich setting. In addition, to maintain a net stoichiometric A/F, the ECM controls the amount of time the injector remains at each setting. An electric fuel pump, which is mounted in the fuel tank, delivers excess fuel to the TBI fuel injector, helping to cool the parts so as to avoid vapor lock. A pressure regulator controls fuel pressure to approximately 70 kPa at the injector.

A special feature of the TBI carburetor is that fuel is injected into the air stream at a specific point so there is only one mixture of fuel and air supplied to the intake manifolds. Therefore, the A/F is always exactly the same for all cylinders of a multi-cylinder engine, regardless of the engine operating condition and any variation in the amount of air delivered to each cylinder via the intake manifold runners.

Fuel Injection to Replace Carburetion

The engineering community at GM realized that carburetors, even at their advanced stage of development in the late 1970s, could not control fuel metering with the accuracy required to operate a three-way catalytic converter. To address this problem, an interdisciplinary team to investigate fuel metering technology was formed under the leadership of George Niepoth of the GM engineering staff.

During many months of intense research and development, which included examining aspects of controls, engine design, and vehicle design, many options were considered and many were discarded. The final solution adopted was a feedback control system (closed loop fuel control), which required an on-board electronic computer to process information and control the timing and amount of fuel supplied to the engine. A sensor in the exhaust stream provided a measure of the exhaust A/F, and the computer drew upon this information to control the amount of fuel supplied, maintaining an A/F at the stoichiometric value, regardless of the power demanded by the vehicle operator.

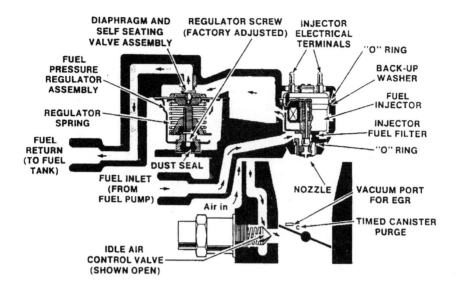

Fig. 6-23 Throttle body injection.

To investigate how to incorporate electronic control of fuel metering and engine operation into a production system, another team was formed under the direction of Robert A. Grimm. This team successfully worked out the myriad details of fuel handling, fuel metering, and engine control logic, as well as the electronics necessary for this novel system. Thus, computer control of the automobile was born! Its arrival changed the basic process of metering fuel for internal combustion engines—a process that had been in use in one form or another since the beginning of the automobile.

The first hardware to implement closed-loop fuel control, used in production for the 1980 model year, was a modified carburetor. The electromechanical carburetor, as it was called, metered fuel using electronically controlled oscillating metering rods. Two years later, the electromechanical carburetor was replaced by a much improved system, throttle body injection (TBI).

The TBI system, by means of electronically controlled injectors, delivered fuel spray directly into the carburetor above the throttle plate. It proved to be an outstanding success and was used throughout General Motors for many years until injectors, intake manifolds, and air meters were developed for manifold fuel-injection systems.

To incorporate this new fuel injection approach into vehicles, the myriad details of fuel handling, feedback control logic, and electronics had to be worked out by GM engineers and technicians. The end result was computer control of fuel metering for an automobile engine, a major breakthrough in the automotive industry. Similar efforts at Ford and Chrysler also led to the development of electronic control of fuel metering for their engines, which in turn resulted in the installation of fuel-injected systems for three-way emission control systems.

Foreign car manufacturers supplying vehicles to the U.S. market also had to develop electronic fuel controls and three-way catalysts. Japanese car manufacturers were approximately in step with the U.S. manufacturers, but European, Asian, and other foreign manufacturers lagged behind.

With a high-speed digital computer electronic control module (ECM) on board, engineers quickly were able to establish computer command controls to govern engine and engine-related vehicle parameters. Table 6-2 lists the range of parameters that were monitored and controlled by the ECM in a typical 1985 vehicle. Without computer controls, the emission levels now being routinely attained would not have been possible.

Table 6-2. Computer Command Control System

Monitored Parameters		Controlled Parameters
Exhaust oxygen concentration Engine coolant temperature Throttle position Barometric pressure Manifold pressure (absolute or differential) Engine crankshaft position Battery voltage Vehicle speed Transmission gear indication Park/neutral mode Brake pedal engagement A/C clutch engagement Time (internally generated in ECM) Cold start program Engine detonation	ELECTRONIC CONTROL MODULE	Carburetor mixture Secondary air valve Secondary air switching valve Electronic spark timing Canister purge valve Torque converter clutch EGR control valve EFE control valve Idle speed control

Legislated emission levels did not change significantly from 1981 to 1993. This period of stability allowed the technical staffs in the automobile industry to refine and optimize emission control systems for improved fuel economy and driveability. These refinements resulted in higher exhaust gas temperatures and more exact fuel control, including fuel shut off during vehicle decelerations. Higher exhaust gas temperatures made it necessary to develop more stable catalyst washcoats to survive high temperatures. When fuel delivery is shut off, the mixture of fuel and air passing through the engine very quickly becomes very lean, requiring a catalyst formulation capable of cycling to the highly oxidizing fuel-off operation without losing conversion efficiency. The catalyst industry responded by using in-house performance and durability bench tests and engine dynamometer tests to improve and refine catalyst and washcoat formulations.

From 1981 until 1990, much R & D in the industry was dedicated to fuel injection technology, and precise control of fuel metering by many alternative fuel injection systems proved to be one of the major accomplishments resulting from these efforts. Throttle body injection, sometimes called single point injection, uses only one injector at the inlet to a manifold. To compensate for the maldistribution of fuel in intake manifold runners, multipoint or multiport injection (MPI) was developed, with an injector at the intake valve of each cylinder. MPI was introduced on several GM vehicles in 1984, followed by tuned-port injection (TPI); and sequential port fuel injection (SPI) in 1985. Ford and Chrysler followed similar steps in introducing their own fuel injection systems.

Sequential fuel injection requires a fuel injector at each intake port, each timed to introduce the exact fuel required just as the intake valve opens. Most engines produced after the mid-1990s employ some version of sequential fuel injection. At the same time that fuel injection was being developed, air meters were designed to measure engine airflow, to ensure that the correct fuel flow could be selected to maintain the correct A/F. In addition, intake manifolds were developed to ensure an even distribution of airflow to each cylinder. Because an automobile engine experiences an overall range of airflow of ten to one, with transient changes in airflow on a millisecond time scale, the importance of accurate airflow measurement cannot be overstatated.

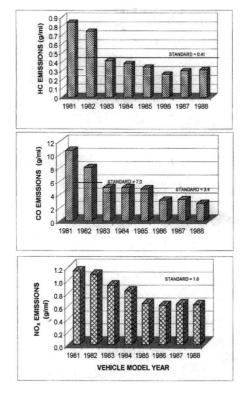

Fig. 6-24 Average 50K emissions for GM passenger cars.

Technical developments between 1981 and 1993 had a great impact on improved vehicle driveability, emissions compliance, and catalyst durability. The use of three-way catalysts, combined with the used of a computer-controlled fuel-delivery system, resulted in a significant lowering of exhaust emissions, especially as for aged vehicles. GM measured the in-use emissions of 4000 vehicles in customer service, including passenger cars and trucks covering model years 1983 through 1988 [6.24]. As shown in Fig. 6-24, test results indicated that there was a steady decline in overall average emissions for GM passenger cars.

In 1990, EPA completed its testing of 833 U.S. vehicles drawn from a sample test fleet for emission compliance. This fleet included vehicles from all U.S. manufacturers manufactured for model years 1983 to 1990. Odometer readings varied from 23,305 to 46,293 miles. Eighty percent of these vehicles passed the HC standard of 0.41 g/mi; sixty-three percent passed the CO standard of 3.4 g/mi; and eighty-eight percent passed the NO_X standard of 1.0 g/mi.

Audit and Customer Inspections

As cars with emission control systems began to be assimilated into the country's overall vehicle population, environmentalists became concerned that production vehicles were not meeting the legislated emission standards. This concern was based in part on environmental measurements of air quality, the results of which indicated that expected air pollution had not decreased as drastically as projected, especially in southern California.

Soon after vehicles with emission control systems appeared in California in 1966, the California Air Resources Board established a vehicle surveillance program to obtain data on the performance of exhaust emission control systems. Vehicles were randomly selected and tested for exhaust emissions, using a duplicate of the test procedures and equipment employed by the auto industry for prototype approval. By 1970, more than 8000 vehicles had been tested.

In 1968, as a result of finding that too many vehicles were not meeting emissions standards, the California legislature required the Air Resources Board to adopt regulations specifying that emission testing of newly produced vehicles be conducted at the end of production lines [6.25]. Results from these end-of-line audits identified specific vehicle models and engine combinations that did not meet the California emissions standards. These findings led California in the mid-1970s to implement selective enforcement audits (SEA). The objective of the SEA program was to identify particular vehicle models and engine combinations that were producing emissions exceeding the appropriate standards.

When several vehicles of a particular model were found to exceed the emission standards, the company that manufactured the vehicle must fix the problem. This required confirming the emissions measurements and tuning the vehicle to the proper settings, which solved most of the problems. In cases where these measures did not fix the problem, the vehicle model was recalled. For some of the recalled vehicles the electronic control module had to be updated, or a component in the emission control system had to be replaced. Occasionally, for vehicles equipped with catalytic converters, it was found that the catalyst had been poisoned by the use of leaded fuel.

Starting in the mid-1970s, EPA patterned its audit program after the California example, employing end-of-line audits for selected vehicle models. EPA began its audits of customer vehicles in the U.S. fleet at the end of the 1970s. EPA also established an inspection program whereby the emission levels of customer vehicles were monitored in specific areas of the United States where air pollution quality was substandard. Each area with substandard air quality, identified as "nonattaining," was required to develop an implementation plan. The objective of the implementation plan was to lower the level of air pollution to satisfy air quality standards. A state implementation program (SIP) might include regulations on fuel availability and restrictions on all sources of air pollution, covering, for example, oil refineries, industrial plants, coal-burning fireplaces, incinerators, and vehicles.

To support the SIPs, EPA in 1981 mandated an inspection program whereby individual states were required to inspect all vehicles licensed within a non-attainment region. Before a vehicle could be registered, it had to pass a local emission inspection test, which included two periods of engine operation at idle and one period at a vehicle speed of 2500 rpm. In this test, the vehicle was not operated on the road or on a chassis dynamometer. CO and HC emissions were measured at the tailpipe and compared with a set of allowable standards, based on the particular model vehicle and engine combination. The cost of this test to the public was not to exceed $15.00. If a vehicle failed, the owner was responsible for having the vehicle repaired and retested. Repairs could not exceed $75.00 for pre-1981 vehicles or $200.00 for post-1981 vehicles. If a vehicle could not pass this test, it was to be retired. Special purpose vehicles and very old vehicles were excluded from the test.

Such emission tests identified a small number of vehicles that had to be either repaired or retired. However, the overall effectiveness of this test has been questioned by many, primarily because testing an engine for emissions during a low-speed idle and a high-speed idle condition does not match vehicle operation on the road. To test vehicles at operating conditions that more closely matched those likely to be encountered during road operation, a second inspection test was developed by EPA; identified as the "EPA Enhanced Inspection and Maintenance test," also known as the "IM 240 test." This test was developed for use in areas with air quality designated as

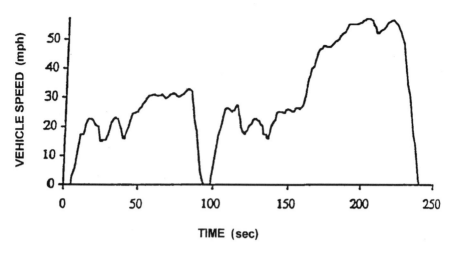

Fig. 6-25 IM240 driving schedule.

"serious," "severe," or "extreme" non-attainment, which included 23 states. Testing began in January 1995 for vehicles and trucks of less than 8500 gross vehicle weight.

The IM 240 test requires that the test vehicle be driven on a chassis dynamometer with real-time measurements of HC, CO, and NO_X. It is known as the IM 240 test because the vehicle is subjected to a driving schedule that is 240 seconds long. This driving schedule was chosen as a shortened simulation of the federal FTP test, as shown in Fig 6-25. The IM 240 test consists of two cycles. The first cycle is the same as the first cycle of the FTP without a cold start. The second cycle starts with the same acceleration as the second cycle of the FTP, but then includes a certain amount of high-speed driving up to 60 mph.

Summary

Beginning in the 1970s, the U.S. auto industry embarked on what would turn out to be one of the most challenging and exciting decades in its history. With the initial impetus coming from California, elected officials at the state level were made to feel the mounting public pressure to control emissions from automobiles, and this quickly found its way to the U.S. Congress and the Environmental Protection Agency. Emissions targets deemed "unobtainable" by the automotive industry pushed car companies to respond with massive R & D efforts. The thermal reactor emerged as a promising technology for solving the problem of controlling emissions, but tests proved that it lacked durability. Subsequently, the focus turned to the catalytic converter as the key component of an emission control system designed to meet the stringent emission standards of 1975. An unanticipated benefit of using the catalytic converter turned out to be that engines could be tuned for higher efficiency and improved driveability.

With the basic technology for implementing a vehicle emissions control system in place, certain problems remained. For example, when customer-owned vehicles equipped with catalytic converters were inspected, some vehicles did not meet the required standards. The result was a period of end-of-line audits, selective enforcement audits, and periodic inspections of in-use vehicles. As further improvements and modifications were made to emission control systems, the United States automotive industry could pause and view with pride the progress it had made in developing a technology that had such an immediate beneficial impact on improving air quality nation-wide. Next on the horizon, however, was the challenge presented by the Revised Clean Air Act of 1990.

References

6.1 Herrin, R., Patterson D.J., and Kadlec, R.H., "Modeling Hydrocarbon Disappearance in Reciprocating-Engine Exhaust," *Combustion Modeling in Reciprocating Engines*, edited by J.N. Mattavi and C.A. Amann, Plenum Press, NY, 1980, pp. 447–482.

6.2 Mondt, J.R., "An Historical Overview of Emission-Control Techniques for Spark-Ignition Engines: Part B-Using Catalytic Converters," ICE-Vol. 8, *History of The Internal Combustion Engine*, Book No. 100294, American Society of Mechanical Engineers, 1989.

6.3 Bishop, R.W. and Nebel, G.J., "Catalytic Oxidation of Automobile Exhaust Gases—An Evaluation of the Houdry Catalyst," Society of Automotive Engineers Annual Meeting, 1959.

6.4 Homfeld, M., Johnson, R.S., and Kolbe, W.H., "The General Motors Catalytic Converter," SAE Paper No. 486D, Society of Automotive Engineers, Warrendale, Pa., 1962.

6.5 Briggs, W.S., "Catalysts and the Automobile—25 Years Later," *Applied Industrial Catalysts*, Vol. 3, Academic Press, Inc., 1984.

6.6 MacGregor, J.R., "Rational Attack on Smog, Today's Major Technical Challenge," SAE Journal, Vol. 74, No. 1, Society of Automotive Engineers, Warrendale, Pa., 1966.

6.7 Schwochert, H.W., "Performance of a Catalytic Converter on Nonleaded Fuel," SAE Paper No. 690503, Society of Automotive Engineers, Warrendale Pa., 1969.

6.8 General Motors, "1972 Report on Progress in Areas of Public Concern," General Motors Corporation, 1972.

6.9 McCabe, L.J. and Koel, W.J., "The Inter-Industry Emission Control Program—Eleven Years of Progress in Automotive Emissions and Fuel Economy Research," SAE SP-431, Society of Automotive Engineers, Warrendale, Pa., 1978.

6.10 Miles, D.L., Faix, L.J., Lyon, H.H, and Niepoth, G.W., "Catalytic Emission Control System Field Test Program," SAE Paper No. 750179, Society of Automotive Engineers, Warrendale, Pa., 1975.

6.11 Taylor, K.C., "Automotive Catalytic Converters," *Proceedings for Conference on Catalytic and Automotive Pollution Control*, Brussels, Belgium, 1986.

6.12 Heck, R.M. and Farrauto, R.J., *Catalytic Air Pollution Control*, Van Nostrand Reinhold, Engelhard Corporation, New York, N.Y., 1995.

6.13 Herod, D.M., Nelson, M.V., and Wang, W.M., "An Engine Dynamometer System for the Measurement of Converter Performance," SAE Paper No. 730557, Society of Automotive Engineers, Warrendale Pa., 1973.

6.14 Casassa, J.P and Beyerlein, D.G., "Engine Dynamometers for the Testing of Catalytic Converter Durability," SAE Paper No. 730558, Society of Automotive Engineers, Warrendale Pa., 1973.

6.15 Church, M.L., Cooper, B.J., and Willson, P.J., "Catalysts in Automobiles, A History," *Automotive Engineering*, Vol. 97, No. 6, June 1998.

6.16 Delphi Automotive Systems, *Catalytic Converter Application Manual*, Delphi Automotive Systems, Energy and Engine Management Systems, 1995.

6.17 Mondt, J.R., "Adapting the Heat and Mass Transfer Analogy to Model Performance of Automotive Catalytic Converters," *J. of Engineering for Power*, ASME Trans. Vol. 109, pp. 109–206, 1987.

6.18 Kuo, J.C., Morgan, C.R., and Lassen, H.G., "Mathematical Modeling of CO and HC Catalytic Converter Systems," SAE Paper No. 710289, IIEC Progress Report, SP-361, Society of Automotive Engineers, Warrendale, Pa., 1971.

6.19 Oldsmobile Division, General Motors Corporation, 1980 Oldsmobile Omega Service Manual, General Motors Corporation, 1980.

6.20 Brisley, R.J. et al., "The Effect of High Temperature Aging on Platinum-Rhodium and Palladium Rhodium Three Way Catalysts," SAE Paper No. 910175, Society of Automotive Engineers, Warrendale, Pa., 1991.

6.21 Sims, G.S. and Johri, S., "Catalyst Performance Study Using Taguchi Methods," SAE Paper No. 881589, Society of Automotive Engineers, Warrendale, Pa., 1988.

6.22 Hammerle, R.H. and Wu, C.H., "Effect of High Temperatures on Three-Way Automotive Catalysts," SAE Paper No. 840549, Society of Automotive Engineers, Warrendale, Pa., 1984.

6.23 Matsunaga, S., Yokota, K., Muraki, H., and Fujitani, Y., "Improvement of Engine Emissions Over Three-Way Catalyst by the Periodic Operations," SAE Paper No. 872098, Society of Automotive Engineers, Warrendale, Pa., 1987.

6.24 Haskew, H.M. and Liberty, T.F., "In-Use Emissions with Today's Closed-Loop Systems," SAE Paper No. 910339, Society of Automotive Engineers, Warrendale, Pa., 1991.

6.25 Cattaneo, A.G. and Bonamassa, F., "California Requirements for Assembly-Line Testing of Vehicle Emissions," SAE Paper No. 700672, Society of Automotive Engineers, Warrendale, Pa., 1972.

Chapter 7

Revised Clean Air Act of 1990 and Exhaust Aftertreatment Systems

Of all the air pollutants related to automotive emissions, ozone has proven to be the most difficult to control to its allowable concentration of 120 parts per billion. Ozone results from a unique and complicated sequence of chemical reactions that occurs when reactive hydrocarbons (VOCs) and oxides of nitrogen (NO_X), are "trapped" in the atmosphere and irradiated by sunlight for several hours. Depending on atmospheric air movement, the ozone may be formed several miles downwind from the source of the reactants. Although some reactive VOCs in the atmosphere result as a natural byproduct of vegetation, additional quantities are emitted in vehicle exhaust, from solvents, and from the chemical and petroleum industries. In contrast, NO_X is produced primarily from the combustion of fossil fuels. Major sources include all types of internal combustion engines and electricity generating stations.

The allowable National Ambient Air Quality Standards (NAAQS) for all air pollutants related to automotive emissions were previously discussed in Chapter 2. These standards are enforced by the Environmental Protection Agency (EPA). In 1970, amendments to the Clean Air Act included control of ozone, with 1975 as the original deadline for meeting the new standards. However, two years after this deadline, the Act was amended to extend compliance to 1982 for some areas, and to 1987 for others.

Despite these new emission standards, in 1987 more than 60 areas still exceeded the NAAQS ozone standards, and, as of 1990, there were 98 areas in violation. These violations resulted because overall atmosphere emission reductions envisioned by the original Clean Air Act had not been met. The intent of the original Clean Air Act of 1965 was to lower ozone by approximately 50%, and it was estimated that to achieve that target, both NO_X and VOCs would have to be lowered by as much as 80%. Although some progress toward these targets had been measured nationwide, the attainment of lowered emissions, especially NO_X, had been well below expectations, in spite of the substantial regulatory programs in effect for 20 years. These findings provided much of the urgency to further revise the Clean Air Act. During 1990, revisions to the Act were vigorously debated in the U.S. Congress by industry, air quality advocates, and regulatory agencies. Finally, in November, President Bush signed the Revised Clean Air Act of 1990 [7.1]. With the purchase American Motors by Chrysler in 1987, only General Motors, Ford, and Chrysler remained to respond to these new regulations.

The Revised Clean Air Act of 1990 contained a significant number of additional regulations, with a target of reducing unburned hydrocarbons, now identified as volatile organic compounds (VOCs), and ozone by 50% from ambient levels present in 1990. This act also included additional requirements concerning State Implementation Plans (SIP), on-board diagnostics (OBD), and use of alternative fuels.

State Implementation Plans required states to monitor levels of carbon monoxide and ozone in the atmosphere. If these levels exceeded national air quality standards, the states were required to present and pursue a plan to achieve these standards. The penalty for not meeting these standards was reduced funding from the federal government.

The on-board diagnostics (OBD) requirement specified electronic monitoring of *all* emission-related devices mounted on a vehicle. Beginning with the 1996 model year, all vehicles sold in the United States had to meet "enhanced on board diagnostics" (OBDII) requirements. Monitored items included: (1) catalyst efficiency, (2) engine misfire, (3) oxygen sensor response, (4) leak from the evaporative system, including gas cap, (5) abnormality of the fuel supply system, (6) function of the EGR valve, (7) function of the secondary air system, and (8) abnormal engine sensor output, such as coolant temperature or manifold pressure. If a particular device failed to function according to a predetermined specification, a malfunction indicator light (MIL) alerted the driver. In response to an MIL, the owner was required to have the failed part either repaired or replaced.

After much dialogue with the auto industry, federal regulators realized that to meet these emission regulations, fuel quality also had to meet new standards. In past years the oil industry had done an excellent job of continuing to improve the quality of gasoline and diesel fuels, but to meet the stringent regulations established by the Clean Air Act of 1990, fuel content was required to meet tighter specifications. Gasoline specifications in the future will place limits on allowable lead content, sulfur content, and possibly phosphorous content in the fuel mix.

The automotive industry is aggressively pursuing the use of alternative fuels to power vehicles. Fuels being considered include compressed natural gas (CNG), liquid petroleum gas (LPG), a blend of 85% methanol with gasoline (M85), and reformulated fuels. An in-depth investigation of the impact of fuel additives on automotive emissions, known as the Auto/Oil Air Quality Improvement Research Program (AQIRP), has recently been completed. Alternative fuels are discussed further in Chapter 9.

The Revised Clean Air Act of 1990 also gave EPA the authority to regulate emissions from both gasoline- and diesel-fueled engines used as power sources in a wide variety of non-automotive applications. These applications included off-road engines, recreation vehicles, motorcycles, lawn mowers, airplanes, marine engines, chainsaws, and heavy-duty equipment, such as rail locomotive diesel engines.

The Revised Clean Air Act specified increasingly stringent emissions, and included a phase-in of Tier I and Tier II emissions for federal vehicles, as shown in Fig. 7-1. Relative to 1990 levels, these emission levels amounted to short-term lowering of both HC and NO_X by 39%, and longer-term lowering of HC by 70%, CO by 50%, and NO_X by 80% relative to 1990 levels. Certification limits were established for both 50,000 and 100,000 mile durability. A limit for CO was included for a 20°F (-6.7°C) cold ambient. In addition, in response to intense lobbying, California passed more stringent standards, especially for unburned HC and for NO_X. These standards, identified as transition low emission vehicles (TLEV), low emission vehicles (LEV), ultra low emission vehicles (ULEV), and zero emission vehicles (ZEV), are shown in Table 7-1. (National low emission vehicles (NLEV) have become a standard in later years.) Meeting these emissions regulations for vehicles presented monumental challenges for all vehicle manufacturers.

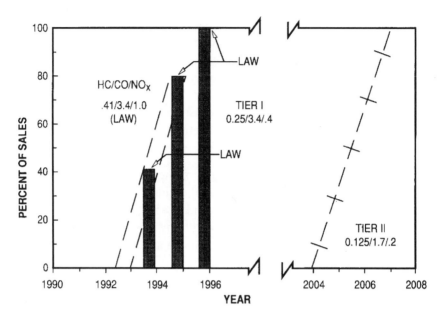

*Fig. 7-1 1990 U.S. Clean Air Act for Passenger Cars and Light Trucks,
Tier I and Tier II Phase-In (<3750 lb).*

Table 7-1. California Emission Standards *
(vehicles < 3750 pounds, 50,000 miles durability)

Note: In this table, non-methane organic gases (NMOG) replaces total hydrocarbons (THC).

	NMOG (g/mi)	CO (g/mi)	NO_x (g/mi)
TLEV	0.125	3.4	0.4
LEV	0.075	3.4	0.2
ULEV	0.04	1.7	0.2
ZEV	0.0	0.0	0.0

* There are many more standards, covering vehicles up to 14,000 pounds and 120,000
miles durability.

The phase-in schedule for NMOG emissions in the United States for both federal and California light-duty vehicles is shown in Fig. 7-2.

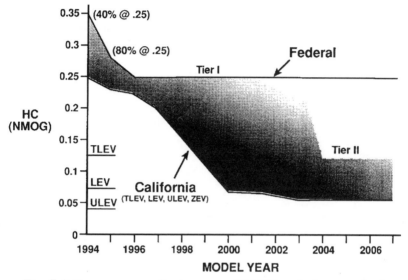

Fig. 7-2 Passenger car fleet average NMOG emission standards.

The California phase-in schedule represented an estimate of the mix of TLEV, LEV, ULEV, and ZEV vehicles that California customers would purchase. The California regulations encouraged automobile companies to meet this phase-in by using financial incentives, to be monitored by California agencies. If the sales-weighted, fleet-average emissions for a given manu-facturer fell below this line, financial credits were awarded. Conversely, if the average rose above that line, financial penalties were to be imposed. According to California law, manufacturers could "trade" these penalties for credits, both within their own product lines and among other manufacturer's product lines.

The catalytic converter, introduced across all vehicle lines in the U.S. in 1975, has proven to be an excellent product to lower exhaust emissions from vehicles powered by spark-ignited engines and burning hydrocarbon fuels. In designing future cost-effective emission control systems into the early 2000s, it is logical that the catalytic converter will remain a primary compo-nent; however, *all* engine and powertrain functions must be controlled as an integrated package to meet these stringent emission requirements.

Since automotive catalysts are very efficient when heated above their light-off temperatures, the most opportune time to control emissions is during engine start-up. Controlling cold-start emissions of HC and CO was not a new challenge. The first systems produced in the early 1970s, prior to catalytic converters, incorporated techniques to lower cold-start emissions (Chapter 5). And techniques to quicken catalyst light-off were part of the early development of catalyst systems (Chapter 6).

By way of illustration, Fig. 7-3 shows the fractions of HC and NO_X emissions discharged in the exhaust of an engine of 1975 vintage, tested using the FTP procedure. From this figure, it is evident that most of the emissions output is from the first (cold start) cycle, but that the hot restart also contributes significant quantities.

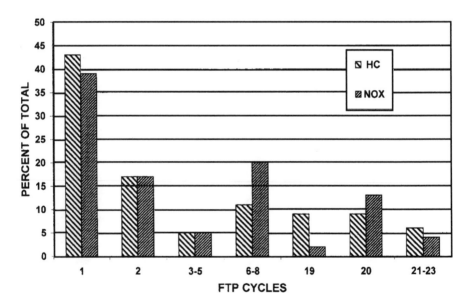

Fig. 7-3 Emissions fractions for a typical LEV on the FTP test.

Based on such test data, and in response to the mandates of the Clean Air Act of 1990, lowering cold-start emissions became the primary focus of industry emissions control development efforts.

Impact of the 1990 Revised Clean Air Act on Exhaust Engineering

With passage of the 1990 Revised Clean Air Act, research efforts were intensified to try to understand better the physics of cold-start emissions in order to size, design, and manufacture catalytic converters. Engineers developing advanced emission control systems immediately realized the importance of either quickly heating the catalytic converter above a threshold reaction temperature of 350°C or, alternatively, developing a catalyst with a much lower threshold temperature for catalyst activity. They reached the conclusion that it would be a much more difficult and time-consuming task to develop a catalyst with a lower threshold temperature for activity than to engineer the exhaust system to quickly heat the catalyst above the threshold temperature for reactions.

Alternative methods for controlling light-off emissions included close-coupled or manifold converters, electrically heated catalysts, and exhaust gas ignition, either through manifold reactions or an in-line burner fired using additional fuel. In addition, a hydrocarbon trap offered the potential to retain unburned hydrocarbons while the exhaust was cold during engine warm-up, and to release these hydrocarbons to be converted by a catalyst when the temperature exceeded the threshold temperature for reactivity.

Exhaust System Thermal Management

Prior to the development of emission controls, heat transfer studies involving exhaust components received little attention. Traditionally, the primary engineering concern relating to heat transfer was to ensure that components in the engine compartment and underbody of the vehicle did not become overheated by radiation and convection from hot exhaust components. However, engineers developing emission control systems soon realized that knowledge about exhaust component heat transfer would be essential in order to optimize the emission control system, especially the operation of the engine.

Conservation of the sensible heat energy available in the exhaust from an engine is necessary to promote gas-phase oxidation reactions in exhaust manifolds or to quickly heat a catalytic converter. To conserve sensible energy and minimize heat loss, auto companies and suppliers aggressively studied thermal energy management of exhaust systems during the cold start period, especially in conjunction with catalytic converters.

A primary objective of using thermal energy management in exhaust systems was to provide a hot catalyst to control cold-start emissions. Two parallel paths were considered: either make catalyst light-off quicker, or keep the catalyst above operating temperature the entire diurnal shut-off time by means of a combination of sophisticated vacuum insulation, heat conservation, and phase change to add thermal energy. Obviously the most economical approach was to utilize the hot gases exiting the engine as the source of thermal energy to heat the catalyst. To evaluate proposed alternative designs, systems engineers combined mathematical modeling with experimentation. This approach proved the least expensive to optimize future exhaust systems.

After the first few firing events, the temperature of the burned exhaust gases exiting the combustion chamber easily exceeded 500°C. However, these hot gases flowed in short pulses as they exited each cylinder, and subsequently were cooled by heat transfer to cold engine exhaust passages and cold channels in the exhaust manifold. It was clear that if the exhaust system could be engineered to minimize the heat transfer processes that cool the exhaust, the converter could be quickly heated shortly after the engine was started.

Many development engineers initially reacted to this challenge by insulating the exhaust passages, especially in the exhaust manifold and takedown pipe leading to the catalytic converter. A variety of insulating materials were sprayed, glued, or mechanically fastened to the inside walls of these exhaust passages. But results were disappointing in that emissions from vehicles with these newly insulated exhaust components did not decrease noticeably during the cold start phase of the US 75FTP.

Meanwhile, much effort was directed at modeling heat transfer in exhaust systems in order to guide the development of heat-conserving systems. The complex physics of exhaust heat transfer processes dictated a combination of analytical and experimental techniques. Mathematical modeling of exhaust heat transfer was based on analysis of a section of circular pipe, one-dimensional in cross section, as shown in Fig. 7-4. Figure 7-4 illustrates the rate equations necessary to account for the disposition of thermal energy, including thermal storage, convective heat transfer, and radiative

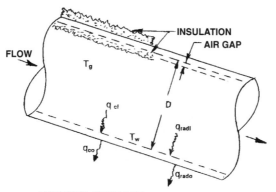

THERMAL STORAGE

$$E_s = Mc_s\Delta T_s = \rho_s\pi DL\delta c_s\Delta T_s$$

CONVECTION

$$q_{ci} = (hA)_i(T_g - T_w)$$

$$q_{co} = (hA)_o(T_w - T_{amb})$$

RADIATION

$$q_{radi} = \varepsilon_i\sigma A_i(T_g^4 - T_w^4)$$

$$q_{rado} = \varepsilon_o\sigma A_i(T_w^4 - T_{amb}^4)$$

Fig. 7-4 Pipe heat transfer model.

heat transfer. Thermal storage in the wall is equal to the mass of the wall, M, multiplied by the product of the wall specific heat, c_s, and the temperature change of the wall, ΔT_s. Convection is the product of the heat transfer coefficient, h, and surface area, A, times the temperature difference at the inside or outside surface of the pipe. Radiation is the surface emissivity, ε, times the Stefan Boltzmann constant, σ, times the surface area multiplied by the fourth power of the absolute temperature difference at the inside or outside of the pipe. To further complicate modeling studies, the processes are time-dependent, both for exhaust gas flow rate, and exhaust gas temperature. Analysis of time-dependent and space-dependent processes results in second order partial differential equations. This usually dictates finite-difference mathematics. Results from one of the pioneering modeling studies was published in 1979 [7.2], with additional studies reported later [7.3, 7.4, 7.5, 7.6, 7.7].

151

A modeling analysis is only as valid as the heat transfer coefficients, and for internal pipe flow in an exhaust system, valid heat transfer coefficients are not easy to obtain. This is because the processes are dynamic and turbulent, including pulsations in flow and pressure and time-variant temperatures. These dynamics result in convective coefficients that do not correlate with proven relationships for turbulent flow, such as those presented by Seider and Tate [7.8].

A follow-up study of convective heat transfer coefficients was undertaken using six exhaust system combinations tested on a 1990 vintage vehicle [7.9]. A detailed study of the convection processes included all exhaust components and resulted in experimental convective coefficients for both the inside and outside of the exhaust components. Internal convective heat transfer coefficients correlated well with the pipe Reynolds number. However experimental results indicated turbulence increased these coefficients by 1.6 for the tailpipe sections to as high as 3.0 for the takedown and manifold sections. Radiation heat transfer accounted for as much as 30% of the steady-state heat transfer from the outer surface of the exhaust components. As much as 50% of the heat loss during steady-state operation from the exhaust components could be charged to the exhaust manifold.

One of the first findings of exhaust thermal energy studies was that a large fraction of the thermal energy available in the exhaust produced by an engine during a cold start is transferred to heat the exhaust manifold. Depending on the specific engine and exhaust manifold, as much as 70% of the exhaust thermal energy exiting the engine during the first 120 seconds of a cold start is transferred to the manifold. This finding prompted much effort to design exhaust manifolds to minimize this heat transfer. Solutions considered included manifolds that were shorter, lighter, or insulated.

Research efforts highlighted two distinctly different functions of insulation materials. The first related to the effects that thermal storage of wall materials had on the light-off characteristics of a catalyst, and the second, to the increase in steady-state temperature resulting from insulating materials.

Studies of the use of insulation to improve thermal energy conservation during a cold-start period revealed that thermal storage of known insulating materials mounted inside an exhaust pipe did not improve thermal energy conservation. Not only did the insulating materials offer no improvement, but they also provided less energy to the converter than a non-insulated bare pipe system (Fig 7-5).

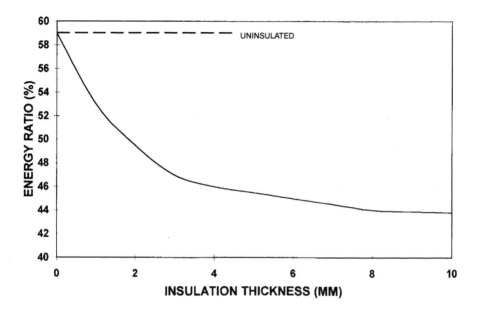

Fig. 7-5 Energy remaining at converter after 100 seconds.
(Source: Ref. [4.7].)

Further studies of alternative exhaust systems demonstrated that the impact of thermal storage in insulated pipes depends on how the particular engine and exhaust system work in combination. An exhaust system receiving relatively cool exhaust temperatures from the engine during warm-up is less sensitive to the storage of thermal energy in the manifold material than an exhaust system exposed to relatively hot exhaust temperatures. A useful technique, when comparing the thermal energy conservation properties of alternative exhaust component designs, is to relate the amount of thermal energy available in the exhaust stream at the converter inlet to that available with a production exhaust system [7.7].

When insulation materials known to have sufficient strength to survive in an exhaust manifold were tested, none was found that would improve the thermal energy available at the converter. The only manifold or pipe design found to have potential was a dual-walled pipe with a thin inner wall (Fig 7-6). As a result of these findings, dual-walled exhaust components with thin inner liners have been developed by both automotive companies [7.10] and suppliers. An obvious concern about a thin-walled inner liner is durability, especially because the thin-walled pipe responds quickly to gas temperature fluctuations as compared with the sluggish thermal response of the outer heavy pipe. Therefore, the design of a thin-walled liner section of pipe must include provisions for mismatch in thermal expansion of the two pipes, as well as provisions to prevent exhaust gas leakage. Leaks can quickly destroy an exhaust system's effectiveness at controlling emissions. Leakage becomes a very critical design issue for flanged joints; it is a concern not only at the joining faces of the flanges, but also at the termination of the pipe and the flange, especially if the pipe is multi-layered.

A number of techniques to conserve heat in exhaust pipes have been reported [7.11]. Multi-layers of insulation, single blankets, and air gaps have been studied at steady conditions and for several cases of engine malfunction.

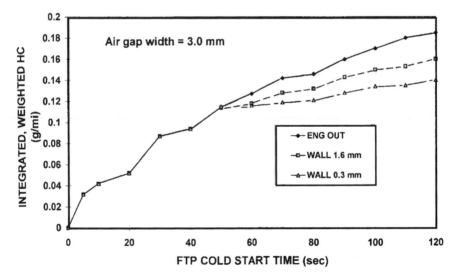

Fig. 7-6 Impact of air-gap inner-liner thickness on cold start emissions. (Source: Ref. [7.13].)

Insulated pipes serve a very important function in shielding other components from hot pipe temperatures, one that is especially important in the engine environment where the trend is toward the use of hotter pipes to control emissions. A variety of design approaches are available, including multiple pipes, multiple layers of insulation, or insulation installed on either the inside or outside of the pipe. All of these design alternatives offer thermal protection for adjacent components, but all increase peak temperatures in the downstream catalytic converter as well. Depending on the specific vehicle, these increases in catalyst temperatures may be as much as several hundred degrees Celsius.

Multidimensional Modeling

Light-off of a catalytic converter departs significantly from one dimensionality because exhaust flow tends to concentrate in the center of a monolith. The higher flow near the center is also the gas with the highest temperature: temperature profiles show that the highest temperatures are in the center, and the cooler temperatures are near the outside surface. These temperature distributions have an impact on emissions performance, and also contribute to thermal strains and stresses in the catalyst substrate.

Multidimensional temperature profiles are more severely distorted during a cold start, because at the beginning of the cold start the converter is uniformly at ambient temperature. As the hot gases impinge on the front face of the monolith, the front sections heat up very quickly, while the rear sections remain at ambient temperature. This disparity occurs because of the fast response characteristics of the catalyst substrate. As time passes, a temperature wave propagates through the substrate. Meanwhile, transient temperature profiles are also generated in the radial direction, partly because of non-uniform exhaust flow, and partly because the outer extremities of the catalytic converter start at ambient temperature. The combination of transient temperature profiles in both the radial and axial directions was studied using a combination of analytical modeling and experiment [7.5, 7.12]. The mathematical complexity of this analysis, especially considering the multidimensional nature of the problem, required the combination of finite element techniques and finite difference techniques.

Newly acquired knowledge about the thermal behavior of exhaust systems enabled alternative designs for exhaust systems to be compared analytically before experimental prototypes were built and tested [7.13]. Modeling results compared the effects of takedown length on cold-start hydrocarbon emissions, and the effects of alternative dual-wall takedown pipes on cold start hydrocarbon emissions. Design tradeoffs for an exhaust system with dual converters were also studied. Fig. 7-7 summarizes results from modeling studies which compare hydrocarbon emissions during a cold start for a dual converter system with those for a single underfloor system. For a dual converter system, these results indicate an approximately 30% reduction in hydrocarbon emissions during the first 120 seconds of an FTP, compared with those for a single underfloor converter system.

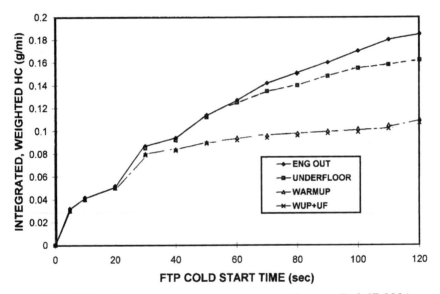

Fig. 7-7 Dual converter system emissions. (Source: Ref. [7.13].)

An unusual study to compare alternative exhaust system designs utilized a statistically designed *modeling* approach [7.14]. Statistical methods are frequently used to design an experimental approach in which several variables are involved. However, even when using statistical methods, small differences in performances cannot always be quantified because of uncertainties in experimental results, such as those caused by hardware variations and measurement variables. In contrast, small differences in performance of alternative systems can be identified using mathematical calculations. Then, an experiment can be designed to confirm the modeling results by testing only one or two exhaust system designs.

In this study, four alternative designs for exhaust systems were compared: (1) inner wall thickness, (2) air gap thickness, (3) takedown pipe length, and (4) pipe diameter, resulting in sixteen combinations of factors. The most important factors contributing to improved emission controls were takedown pipe length and the thickness of the inner wall for a dual-wall design. An air-gap width of at least two mm (0.08 in.) was shown to be necessary for good performance of a dual-pipe design.

Following up the modeling studies, two sets of hardware were built and tested on a vehicle. One system used a single-walled pipe, while the other used a dual-walled pipe. Test results confirmed the findings of the initial modeling study that, for quick catalyst light-off, an exhaust system should use a dual-walled pipe with thin inner pipe walls, a short takedown length, and small overall exhaust pipe diameter. The dual-wall pipe also was found to increase converter bed temperatures during steady-state operation of the vehicle by approximately 200°C at a vehicle speed of 60 mph.

Three-Way Converter Modeling

For a three-way catalyst to effectively control HC, CO, and NO_X emissions, the A/F must be maintained very close to the stoichiometric value. A deviation limit of only ± 0.2 A/F about the stoichiometric A/F has been identified through experimental studies to be the optimum target value. The need to maintain this precision in A/F forced the auto industry to develop a feedback electronic-control system for fuel metering. Feedback control was accomplished by cycling the A/F between rich and lean fueling limits.

To accommodate this A/F cycling, catalyst additives have been developed to augment conversion performance by storing and releasing oxygen during each cycle. Thus, catalysts for three-way operation to convert NMOG, CO, and NO_X operate continuously, cycling from lean-to-rich and rich-to-lean A/F, resulting in a time-averaged conversion efficiency for any one constituent. The continuous cycling of A/F from rich-to-lean-to-rich about the stoichiometric value produces a "net" stoichiometric A/F, at which conversion efficiencies for both reducing and oxidizing reactions are less than the maximum possible—

a factor that adds to the already complex nature of catalysis in an automotive exhaust, with its transients in flow, constituent concentrations, and temperatures. Consequently a complete closed-form analysis of the performance of a catalyst is not currently possible. However, modeling in which the empirical relationships are skillfully combined to simplify the analysis has proven useful [7.15].

Modeling can be useful in understanding critical behavior trends in emission control. In turn, these trends are used to optimize parameters in the emission control system. Obviously, the performance of a catalyst is strongly dependent on the operation of the engine, and on engine control parameters such as A/F dynamics, spark advance, and EGR controls. The ultimate objective of modeling efforts is to help the emissions control development engineer   compare alternative parameters that have an impact on catalyst size and chemical characteristics. Those results would provide a rationale for designing a catalytic converter that is of the appropriate size, and has the necessary thermal, physical, and chemical characteristics, to meet the legislated requirements for emissions. As suggested in Fig. 7-8, if modeling can calculate conversion efficiencies as a function of converter size, then a sizing correlation can be used to identify the necessary converter superficial surface area required for a given conversion performance.

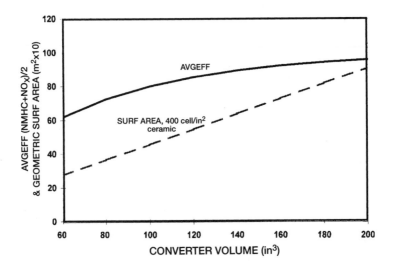

Fig. 7-8 Typical NMHC and NO$_X$ efficiency.

Thermal Durability

The thermal durability of the catalyst and substrate dictate much of the design of an exhaust system. If the catalyst temperature is controlled between approximately 400 and 800°C, a much more durable catalytic converter, occupying less volume, can be designed for a specific vehicle. If operating temperatures exceed 800°C, either the durable life of the catalyst will be limited or a more expensive substrate and washcoat combination will be required (as discussed in Chapter 6). One solution to the problem of operating at high temperature is to use a catalytic converter of larger volume.

When a catalytic converter is added to the exhaust system of a carbureted engine, high-speed engine overrun, or vehicle coast down, during which the throttle is closed but the engine continues to rotate, can cause overheating of the catalyst. Depending on the severity of the overrun, the closed throttle produces a vacuum in the intake manifold, which pumps gasoline into the engine; however, combustion of this gasoline is not required to propel the vehicle, and misfiring results. In this way, a combustible mixture of air and fuel reaches the catalytic converter. A fully warm catalytic converter will reliably promote the exothermic oxidizing reactions of the unburned fuel. When excessive unburned fuel is supplied to an operating catalytic converter, it is possible for temperatures to reach levels high enough to cause physical damage to the substrate.

Early modifications to engine controls to minimize misfiring during coast down included higher-voltage ignition to improve reliability in igniting the charge and setting the throttle stop so the throttle could not be completely closed. A disadvantage of not allowing the throttle to close completely was "after-run" of the vehicle after a normal stop, when the operator turned off the ignition. This occurred because ignition interrupt does not stop fuel flow, and with the throttle blocked partially open a mixture of fuel and air was supplied to the combustion space. If the combustion chamber contained a hot spot, such as a carbon deposit, this fuel would be autoignited, causing the engine to continue to run.

A design alternative to minimize overheating resulting from engine overruns was a "*guard*" catalyst system. The guard catalysts were two small convert-ers located in the exhaust manifold, upstream from the main underfloor catalytic converter. They were designed small so that only a fraction of the incoming unburned hydrocarbons or carbon monoxide could be oxidized, not enough to damage the guard catalysts. This system was demonstrated to be successful in a closed-throttle coast test [7.16]. The guard catalysts lowered the maximum temperature of the main converter by up to 230°C, with no significant damage to the guard catalysts. The guard catalyst concept was not used in production because engine modifications proved to be sufficient to control overheating excursions to acceptable levels. However, it provided some of the early technology for close-mounted catalyst design.

At the time when catalytic converters were first installed on vehicles with carburetors, there was no mechanism available to shut off the fuel supply to the engine during an engine overrun or vehicle coast condition. What was required to accomplish fuel shut off was a technology capable of sensing with millisecond response engine speed, throttle position, and manifold vacuum, and of being mated with a fuel metering system with the same response. Fuel injection systems fulfilled these requirements; one of its features was the ability to shut off the fuel supply during an overrun and restart fueling with no impact on vehicle driveability. As fuel injection was incorporated into engines, so was fuel shut-off during coast downs. This essentially eliminated this source of overheating in catalytic converters, and simultaneously improved vehicle mileage.

Alternative Subsystems

Alternative subsystems to make catalyst light-off quicker can be divided into two subcategories: (1) passive, with no moving components, and (2) active, containing moving components. Passive subsystems include thermal energy conservation systems, close-mounted converters, warm-up converters, and a simple adsorber. Active subsystems include warm-up converters with valves, supplemental heat systems, and adsorbers with valves. The relative effectiveness of alternative techniques [7.17] is compared in Fig. 7-9. Obviously, active techniques are more complex, with attendant higher cost and lowered reliability.

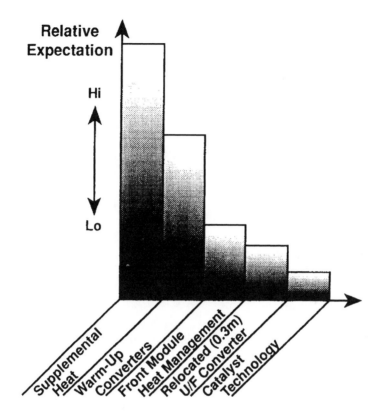

Fig. 7-9 Technologies for lowering cold start emissions.

Passive Subsystems

The simplest, most economical approach to make catalyst light-off quicker, by retaining sensible heat in the exhaust from an engine, is to add insulation (Fig. 7-10). However, making room for the insulation is a packaging concern. Further, the retained heat will cause the converter to operate at a higher temperature, with the attendant risk of accelerated thermal degradation. Dual-wall pipes are a design option in which an air gap is the insulating medium. Limited results from several vehicles showed that dual-wall downpipes generally lowered emissions of HC, CO and NO_X. These results are shown in Table 7-2.

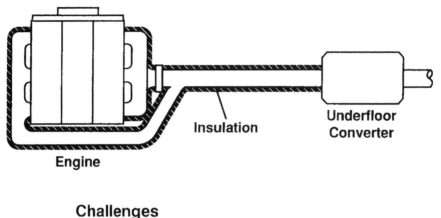

Challenges
- **Catalyst Thermal Aging**
- **Reliability, Durability**
- **Space**
- **Cost**

Fig. 7-10 Underfloor subsystem with thermal energy conservation.

Table 7-2. Measured Tailpipe Emissions for Dual Wall Downpipes
[(-) denotes improvement]

Constituent	Average Decrease (%)	Smallest Decrease (%)	Largest Decrease (%)
HC	-21	-6	-33
CO	-9	+5	-23
NO_X	-12	+14	-43

The most effective technology for quickly heating a catalytic converter is to mount the converter at or very near the exhaust manifold. Installation at this location permits very rapid heating of the catalyst early in the cold start, to convert exhaust pollutants [7.18]. A drawback of close mounting is that higher catalyst temperatures are generated. Many companies, especially Ford, Honda, and Toyota, have chosen to add warm-up converters mounted close to the exhaust manifold. These warm-up converters heat up quickly because of their close proximity to hot exhaust gases discharged from the engine. However, packaging is a challenge, especially in a V-6 installation, as shown in Fig. 7-11. Furthermore, warm-up converters suffer accelerated thermal degradation during high-speed, high-load operation of the engine. Studies have been done of the feasibility of adding a valve to bypass exhaust gas past a warm-up converter except at start up; however, to date, no valve has been designed that provides low leakage, long life, and acceptable cost.

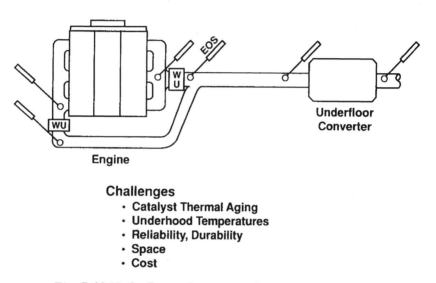

Challenges
- **Catalyst Thermal Aging**
- **Underhood Temperatures**
- **Reliability, Durability**
- **Space**
- **Cost**

Fig. 7-11 Underfloor subsystem with warm-up converters.

To monitor the performance required for OBD (on-board diagnostic) compliance, oxygen sensors may be required before and after each converter. Thus, for a V-6 or V-8 installation with warm-up converters, six exhaust oxygen sensors (EOS) would be required, along with the accompanying wiring and electronics.

In order to fit them in the engine compartment, warm-up converters are generally made smaller than underfloor converters. A warm-up converter, often referred to as a "pup," warms up much more rapidly than an underfloor converter, but suffers greater thermal deterioration. To satisfy durability requirements, a warm-up converter requires a "high-temperature" catalyst substrate and washcoat. To minimize converter exterior wall temperatures, which could thermally damage adjacent components in the engine compartment, a warm-up converter usually incorporates built-in insulation around the substrate and in the end cones.

Thermal energy storage in the warm-up converter slightly delays heating time of the underfloor converter. However, in the long term, the warm-up converter protects the underfloor converter from "spikes" in emissions, which extends the life of the underfloor converter. Overall, emission conversion performance during aging for a combination of warm-up and underfloor converters has proven to be quite good.

Vehicle Emission Results

Results from a study using 163 in.^{3} (2.67 L) of total catalyst volume to control emissions from a 1991, 2.3-L four-cylinder engine have been reported [7.19]. The total converter volume was allocated in different proportions between the warm-up and underfloor locations. All converters were aged using an accelerated engine schedule on an engine dynamometer to the equivalent of 50,000 miles of U.S. driving. The results of these tests, which are presented in Fig 7-12, indicated that the most effective use of catalyst volume is to divide the total volume into a small-volume warm-up converter and large-volume underfloor converter. Results from this study indicated that TLEV emission levels could be achieved for this vehicle with a warm-up and underfloor system.

If space permits, larger close-mounted converters may be used, especially in large cars and trucks. The major advantage of using one large close-mounted converter is simply fewer converter components to fabricate and package. The combination of a small warm-up converter directly attached to a larger converter is known as a "cascaded" converter [7.20]. This concept permits higher-temperature stabilized catalysts, such as palladium, to be used in the small converter and less expensive catalyst formulations to be used in the larger converter.

Attempts have been made to mount a catalyst directly in the exhaust manifold to determine its potential to meet ULEV emissions. Results have proven poor durability with this arrangement because a catalyst mounted in the exhaust manifold is exposed to high temperatures and flow pulsations.

Constant Total Volume

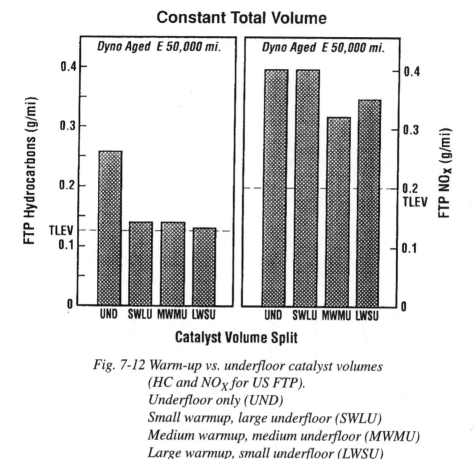

Fig. 7-12 Warm-up vs. underfloor catalyst volumes
(HC and NO_X for US FTP).
Underfloor only (UND)
Small warmup, large underfloor (SWLU)
Medium warmup, medium underfloor (MWMU)
Large warmup, small underfloor (LWSU)

Simple Adsorber

A "simple" adsorber system represents a promising technology for storing unburned hydrocarbons during a cold start, until the catalyst reaches light-off temperature (Fig. 7-13). After light-off, the higher temperatures from the engine cause the hydrocarbons to desorb from the adsorber and react on the hotter catalyst surface.

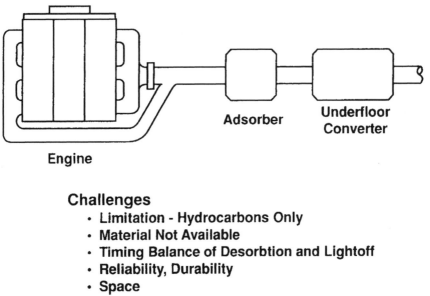

Adsorber　**Underfloor Converter**

Engine

Challenges
- **Limitation - Hydrocarbons Only**
- **Material Not Available**
- **Timing Balance of Desorbtion and Lightoff**
- **Reliability, Durability**
- **Space**
- **Cost**

Fig. 7-13 Underfloor subsystem with simple adsorber.

Unfortunately, zeolites or carbon materials, both materials can adsorb hydrocarbons, are unable to tolerate the temperature produced in the exhaust of current internal combustion engines. However, adsorbers might be more practical for diesel engines, in which the temperature of the exhaust is much lower.

To function successfully, an adsorber material must adsorb at room temperature; be a "willing" desorber, allowing the adsorbed hydrocarbons to be purged at catalyst temperatures; and tolerate high temperatures during fully warm operation of the engine. To accommodate the multitude of vehicle operating conditions, valves will most likely be required to control flows and temperatures in an automobile adsorber system. At this time, evaluation of simple adsorber systems indicates that approximately 50% of the cold-start hydrocarbons can be successfully adsorbed on materials available [7.21]. Although adsorber technology is still in its infancy, and involves many variables, in combination with catalytic converters it offers much promise as an important component of future emission control systems.

Active Subsystems

Prior to or immediately after engine start, supplemental heat can be supplied to a catalyst by either electrical or fuel combustion energy. This provides a hot catalyst, to control cold-start emissions to very low levels. Heated catalyst systems were the most favored technology to attain ULEV emissions of 0.04, 1.7, and 0.20 g/mi for NMOG, CO, and NO$_X$, respectively. In the early 1990s, development of an electrically heated converter (EHC) system, shown schematically in Fig. 7-14, was very actively pursued by the auto industry, suppliers to the auto industry, and private laboratories [7.22].

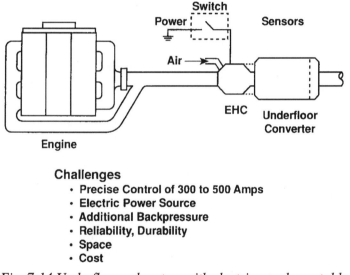

Challenges
- Precise Control of 300 to 500 Amps
- Electric Power Source
- Additional Backpressure
- Reliability, Durability
- Space
- Cost

Fig. 7-14 Underfloor subsystem with electric supplemental heat.

Two years earlier, in 1988, the EPA and the California Air Resources Board (CARB) jointly funded initial development studies at Southwest Research Institute (SWRI) [7.23]. Hand-built, experimental EHC systems were incorporated into several modified production vehicles at SWR. Greatly lowered emissions were obtained for these vehicles by preheating a monolith catalyst prior to starting the engine. By having the catalyst above light-off temperature, the cold start hydrocarbons, as compared with emissions from a standard vehicle, were significantly lowered. These results were a significant factor in establishing the California emission levels shown in Table 7-1.

Components of an electrically heated converter system include a battery or alternator power source, control system, modulated air injection, and sensors. Although EHC systems operate for only a few seconds, at least 2 kilowatts of energy from a 12-volt battery source is required. This level of current flow dictates the use of high-capacity wiring and expensive switching equipment to control 200 to 300 amperes. And in the process of triggering the EHC, the deep discharge must be monitored, and interrupted if necessary, so as to not interfere with engine starting.

Mounting the EHC adjacent to the engine lowers the amount of energy required, but the close location subjects the catalyst to higher temperatures during all fully warm operating conditions of the engine. The higher operating temperatures, in turn, cause more rapid thermal aging of the catalyst. Another drawback of the EHC system is that energy to charge the batteries detracts from overall vehicle fuel economy.

Battery technology necessary for the deep discharge required at every cold start is beyond the current production standards for an automotive battery. In addition to improved battery technology, a significant challenge to emission control engineers is to design and build a catalytic EHC that can survive in the exhaust system environment. Such an EHC must be robust to the flow of hot, pulsating exhaust gas, and must also be able to stand up to vehicle vibrations and contamination from water and debris splashed from road surfaces.

Southwest Research Institute [7.23] and Corning Inc. [7.24] have reported emission levels below ULEV targets for EHC systems. Other results from EHC development efforts, including many vehicle installations, have been reported in the literature by the U.S. Environmental Protection Agency [7.25]; W.R. Grace & Co. [7.26]; Camet [7.27]; and Corning [7.28]. To assess the state of development by the end of 1991, several organizations prepared summary reports of emissions measurements taken from vehicles that they had built and tested [7.29]. These publications summarized the measurements of tailpipe emissions for unburned hydrocarbons (Fig. 7-15) and for oxides of nitrogen (Fig. 7-16). Figures 7-15 and 7-16 summarize data for a range of vehicles.

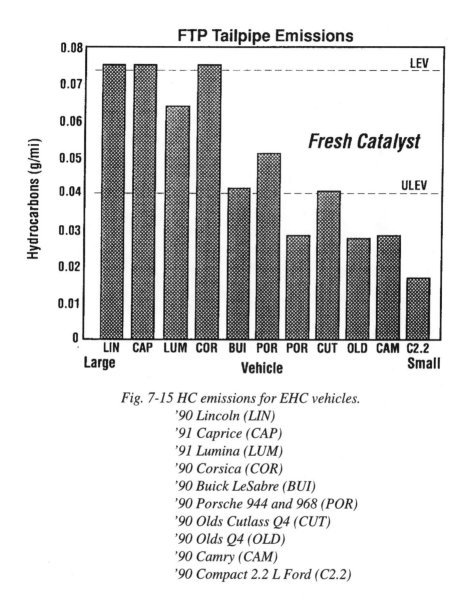

Fig. 7-15 HC emissions for EHC vehicles.
'90 Lincoln (LIN)
'91 Caprice (CAP)
'91 Lumina (LUM)
'90 Corsica (COR)
'90 Buick LeSabre (BUI)
'90 Porsche 944 and 968 (POR)
'90 Olds Cutlass Q4 (CUT)
'90 Olds Q4 (OLD)
'90 Camry (CAM)
'90 Compact 2.2 L Ford (C2.2)

From these figures it is evident that EHC systems on large vehicles with large engines barely met the LEV emissions levels, while EHC systems on smaller vehicles with smaller engines barely met ULEV emissions levels. For oxides of nitrogen, the larger vehicles failed to meet either LEV or ULEV emissions targets, while the smaller vehicles barely met both LEV and ULEV emissions targets.

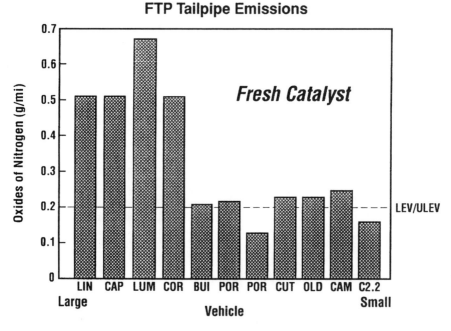

Fig. 7-16 NO_X emissions for EHC vehicles.

Emissions from a Honda EHC system were reported to be as low as 0.03, 0.34, and 0.07 g/mi for the US 75 FTP for NMOG, CO, and NO_X, respectively, with the development vehicle powered by an L-4, 2-L, engine [7.30]. The first production vehicle using an EHC system was a BMW powered by a V-12, 5.7-L limousine engine [7.31]. This vehicle has a unique system with one EHC on each manifold and alternate heating of the converters to minimize electric current flow from the alternator. Durability for this BMW system is promising, based on service in customer use through mileage up to 40,000 kilometers. These results demonstrate that EHC converters from all manufacturers are being designed with improved durability [7.32].

EHC systems were also tested on trucks. In the U.S., emission standards for trucks up to 14000 lb GVW were phased-in beginning in 1997, and those for trucks greater than 14000 lb GVW, beginning in 1998. Since the duty cycle for a typical truck is more severe than that for a typical passenger car, engineering EHC durability for a truck emission control system poses an additional challenge.

An alternative to EHC is a supplemental heat system using an auxiliary fuel, usually gasoline, and a burner to provide thermal energy to heat a cold catalyst. From an energy perspective, the fuel-fired burner offers much more efficient use of energy than does an electrically heated system. As shown in Fig. 7-17, for one unit of energy available in fuel, potentially 95% can be converted to thermal energy to heat a catalyst.

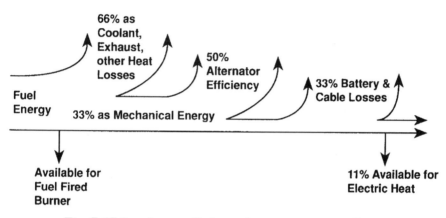

Fig. 7-17 Supplementally heated converter energy flows.

To obtain the electric energy for the EHC, this same unit of fuel energy must be converted to shaft work using the vehicle engine. Next, a belt-driven alternator is used to generate the electric energy. By this time, the available energy for the EHC is probably lowered to about 16% of the original fuel energy produced. After energy losses for controls, cables, and connectors are subtracted, the result is that only about 11% of the original fuel energy is available to heat the catalyst.

A fuel-fired burner system to heat a catalytic converter requires a combustion air pump, fuel supply, ignitor, and control system, as shown in Fig. 7-18. A schematic for such a system, developed and demonstrated on both cars and trucks, is shown in Fig. 7-19. This system is somewhat complex; it requires an efficient electric-driven air blower and no-leak check valve at the blower exit to block raw emissions from backflow after the blower is shut off. This and other fuel-fired burner systems have demonstrated the potential to control HC to ULEV levels for passenger cars with weight less than

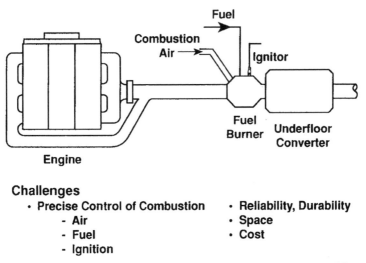

Challenges
- Precise Control of Combustion
  - Air
  - Fuel
  - Ignition
- Reliability, Durability
- Space
- Cost

Fig. 7-18 Underfloor subsystem with fuel supplemental heat.

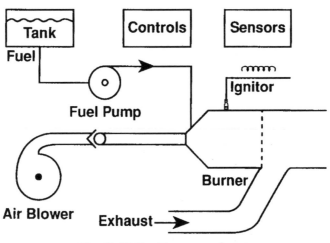

Fig. 7-19 Fuel burner subsystem.

3750 pounds gross vehicle weight (GVW), as shown in Fig. 7-20. But the ability to meet ULEV NO_X is not quite as good, as shown in Fig. 7-21, where the range of uncertainty in the measured data shows some test results exceeding the target standards. As a result, emissions performance is much the same as that of EHC.

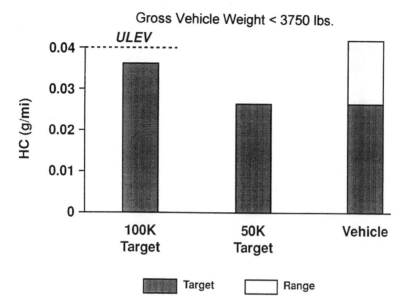

Fig. 7-20 Vehicle FTP HC emissions.

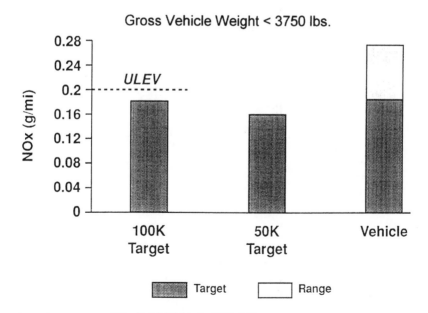

Fig. 7-21 Vehicle FTP NO$_X$ emissions.

There are pros and cons for both EHC and burner supplemental heat. On balance, the EHC system is simpler. However, the fuel-fired system offers more packaging flexibility, and is probably more robust with regard to long-term durability.

A simple adsorber system does not seem practical because adsorber materials available to date have limited maximum temperature tolerance. To protect an adsorber from damage by overheating, one or more valves can be used to purge the adsorbed HC. A typical underfloor system with control valves is shown schematically in Fig. 7-22. In this system, during the vehicle starting sequence, the main valve directs exhaust flow through the adsorber to "collect" unburned hydrocarbons when the exhaust is cool. When the catalytic converter is heated above light-off temperature, the valve directs flow through the underfloor converter. The valve can be modulated, or another valve can be included to purge the hydrocarbons from the adsorber into the intake manifold, using manifold vacuum as the driving pressure differential. Prototype vehicle adsorber systems were assembled and tested by many investigators in the early 1990s.

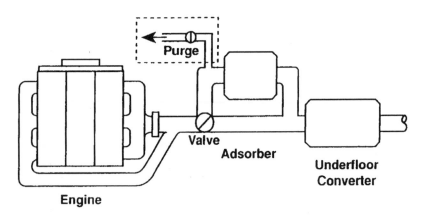

Challenges

- Limitations - Hydrocarbons Only
- Material
- Purge
- Reliability, Durability
- Space
- Cost

Fig. 7-22 Underfloor subsystem with adsorber and valve

A sample of a valved system built and tested at Southwest Research is shown in Fig. 7-23. A 60 in.^{3} (0.98 L)"burn-off" catalyst and 60 in.^{3} (0.98 L) adsorber were added downstream from the production 170 in.^{3} (2.80 L) catalytic converter; and a three-valve purge system was used to control exhaust flow. Using adsorber materials provided by several suppliers, vehicle tailpipe emissions for this vehicle were measured on the 75 FTP. For four candidate adsorbers, tailpipe NMHC varied from 0.045 to 069 g/mi; CO varied from 0.638 to 0.805 g/mi; and NO$_X$ varied from 0.133 to 0.16 g/mi. These emissions measurements for experimental prototype systems indicate the potential to meet LEV emission standards for NMHC (non-methane hydrocarbons) and CO, and ULEV standards for NO$_X$.

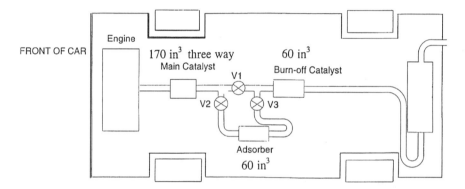

Fig. 7-23 Valved adsorber system. (Courtesy of Southwest Research Institute.)

An exhaust system scenario chart, Fig. 7-24, shows generic applications of passive and active techniques for various vehicle "sizes" and various emission levels. Emissions standards are correlated as a function of vehicle emission categories: small, medium, and large. These categories are part of a general classification scheme combining engine size and vehicle size. Passive systems are indicated for small categories through TLEV emissions levels and for some LEV levels. Passive subsystems are appropriate for small and medium vehicles for TLEV and a few for LEV. Passive systems for large vehicles will only meet Federal Tier I and TLEV requirements. Active systems will be developed for small ULEV and medium LEV requirements. ULEV emissions levels for large vehicle categories dictate complex systems, most likely combining several technologies.

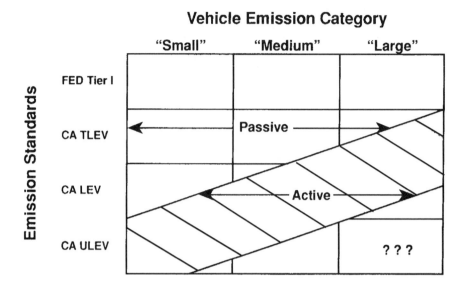

Fig. 7-24 Future exhaust subsystems scenarios.

Exhaust System Pressures and Pressure Drop

The engine provides the pumping power necessary to force exhaust flow through exhaust components to overcome the pressure loss through each component. As exhaust systems become more sophisticated, the impact of exhaust component pressure drop on engine power becomes more critical. This is especially true as vehicles and engines are continually updated to improve fuel economy. Improved fuel economy is important because it provides a financial benefit to the customer, reduces use of gasoline fuels, and lowers the amount of greenhouse gases (e.g., carbon dioxide) released into the atmosphere.

The effect of exhaust "backpressure" on power for a typical engine is shown in Fig. 7-25. The amount of power loss depends on many factors, but a good "rule-of-thumb" is that one inch (25.4 mm) of mercury backpressure causes approximately 1.0% loss of maximum engine power. Although the laws of physics require some "pumping power" to discharge the exhaust, some control of this pressure drop is possible through adjustment of geometric factors, design options, and manufacturing process controls. To intelligently allocate exhaust pumping power requires detailed knowledge of the sources and magnitudes of pressure drop components.

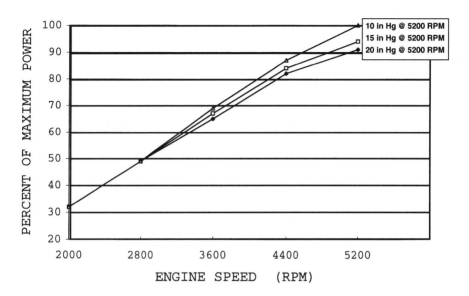

Fig. 7-25 Effect of exhaust backpressure on power.

Much of the information on pressure drops that is available in the literature has been generated for non-automotive applications in which flow is steady without pulsations. In addition, pressure drops for internal pipe flow have been studied thoroughly for steady turbulent flow. However, the turbulence levels for pressure drop correlations reported in the literature are not nearly as severe as those experienced in the exhaust manifold and takedown pipes in automotive exhaust systems.

Flow Processes

Flow inside a duct, whether laminar or turbulent, is characterized by Bernoulli's equation (Eq. 1), which relates flow velocity head, or dynamic head, to static and total pressure [7.33]:

$$P_{tot} = P_{static} + \rho V^2 / 2 \tag{1}$$

$$\text{velocity head}$$

Velocity head has dimensions of pressure, and is easily calculated for exhaust flow if the flow rate and gas density are known. Values of velocity head for the exhaust from a typical engine at wide-open throttle flow are presented in Fig. 7-26. This graph shows that a velocity head of 1.0 in. Hg. is possible for small exhaust pipes.

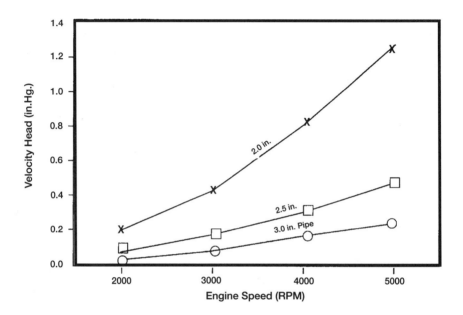

Fig. 7-26 Exhaust flow velocity head.

Relationships between velocity head, total pressure, and static pressure for a frictionless duct with area change are depicted in Fig. 7-27. Small changes in pipe diameter have a large influence on velocity head, which varies inversely as the fourth power of diameter.

Also illustrated is the risk of measuring pressure loss in a pipe system if pipe sizes change. For the system illustrated, static pressure taps would indicate an increase in pressure or negative pressure loss for flow in the pipe section.

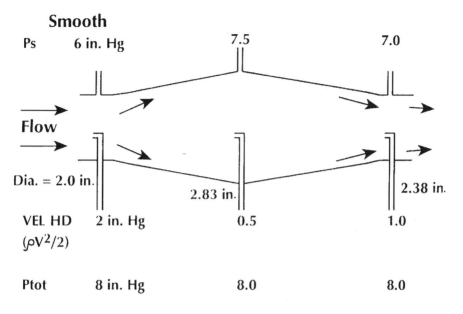

Fig. 7-27 Flow with area change.

Flow friction causes irreversible pressure losses, which can be correlated using Eq. 2 for internal flow in a pipe:

$$\Delta P = 4fL/Dx(\rho V^2/2) \tag{2}$$

As defined by Eq. 2, friction factor can be derived for simple geometries. Thus, values have been measured by many investigators and published in the open literature. However, very little of the reported data is of value for analyzing an exhaust system. This is because experiments to date have not included pulsating flow and partially developed boundary layers, two conditions that exist in an exhaust system. Since pulsating flows augment heat transfer coefficients by as much as a factor of three [7.9], the Colburn analogy (a proven correlation relating heat transfer and friction factors for turbulent flows) would predict a similar increase in friction factors.

Sources of Pressure Drop

In an exhaust system, pressure drop is caused by all components: exhaust manifold, downpipe, junction, converter, resonator, muffler, and each section of pipe. For the catalytic converter, the essential components of pressure drop are the entrance header or inlet cone, exit header or exit cone, and the substrate [7.34]. Total pressure drop is the sum of the individual contributions from these components, with a significant contribution usually provided by the inlet cone because of maldistribution of flow through the substrate. A typical converter pressure distribution is presented in Fig. 7-28. This graph shows that at maximum flow approximately 30% of the total pressure drop is generated in the inlet cone. On the other hand, less than 3% can be charged to the inlet cone for a low-flow road-load operating condition.

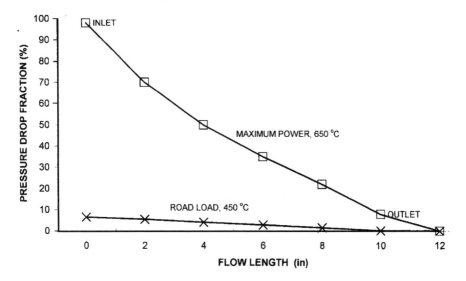

Fig. 7-28 Converter pressure distribution

Substrate

To control pressure drop, it is important that the emissions control engineer be able to modify the geometry of the catalytic converter substrate. Possible ways of doing this include: (1) using alternative substrates, (2) changing the cell density of a given substrate, (3) changing the shape of a given substrate, and (4) changing the inlet and/or outlet pipe size.

Monolith converters have employed a variety of flow-passage cross sections, including square-cell ceramic, plate-fin metal, and herringbone metal. Pressure drop for any monolith substrate is influenced by cell density, wall thickness, and washcoat loading. With all flow passage geometries, an increase in cell density results in more surface area, so pressure drop is increased, as shown in Fig 7-29.

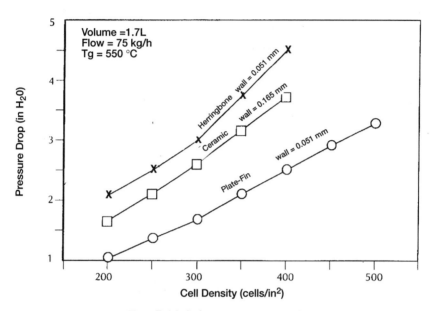

Fig. 7-29 Substrate pressure drop.

Changing the shape of a given substrate can be correlated with changes in pressure drop. This is because substrate geometries essentially operate in a laminar flow regime, such that friction factor is equal to a constant divided by the Reynolds number, as shown in Eq. 3 (symbols are identified at the end of this chapter).

$$f_{laminar} = CONST / Re \qquad (3)$$

then:

$$\Delta P \approx (wL) / A_{fr} \qquad (4)$$

From Eq. 4, substrate pressure drop is essentially proportional to L/A_{fr}, assuming that entrance and exit effects are very small. Therefore, a convenient correlation for comparing alternative converters is pressure drop vs. L/A_{fr}, for a constant flow rate, w. Using two different inlet pipe sizes, as shown in Fig. 7-30, an analytic prediction shows that a near-linear correlation results for a 160 in.^{3} (2.62 L) volume of 400 cell per square inch (62.0 cell/cm^2) ceramic substrate. Departures from a linear correlation would only be possible if the inlet and outlet headers made large contributions to total pressure drop. Note the significance of pipe size on the overall pressure drop.

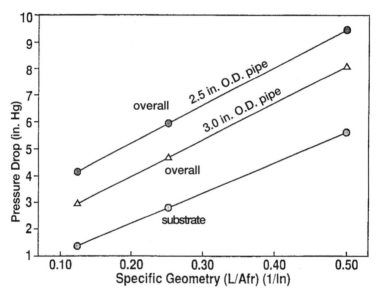

Fig. 7-30 Overall converter and substrate pressure drops.

Inlet Header (Cone) and Maldistribution

A significant increase in pressure drop can result when the flow is expanded from the inlet pipe to the catalytic converter frontal area. Expanding the flow at the inlet causes more additional pressure drop than contracting the flow in the exit header. Flow maldistribution at the substrate inlet face depends on the flow velocity, described as the "velocity head," and the magnitude of substrate pressure drop. As shown in Fig. 7-31, if the substrate flow pressure drop is large, less flow maldistribution results.

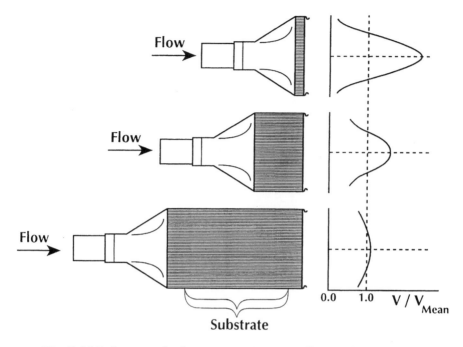

Fig. 7-31 Influence of substrate restriction on flow maldistribution.

Since sufficient pressure drop is required to force flow at the maximum velocity through the substrate, pumping power required to overcome the restriction depends on the maximum velocity through the substrate. Consequently, in correlating the pressure drop caused by flow maldistribution, the "effective" pressure drop is found to depend on the maximum velocity. A correlation of the maximum velocity relative to the mean velocity as a function of the ratio of substrate pressure drop to the velocity head in the inlet pipe presented in Fig. 7-32. This correlation is based on experimental data [7.34]. For a typical catalytic converter, the ratio of substrate pressure drop to pipe velocity head is 5 to 15, values that correspond to logarithmic values of 0.70 and 1.18, respectively, on the x-axis in Fig. 7-32.

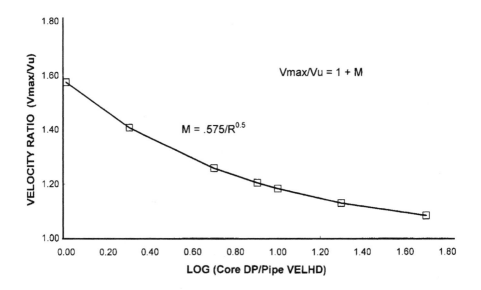

Fig. 7-32 Maldistribution velocity ratio.

Loss Coefficients

For many years, a "loss coefficient," K, has been used to correlate flow losses for flow around bends and through abrupt expansions and contractions, Y-joints, and flow dividers. K is defined in terms of the inlet velocity head, as shown in Eq. 5, and accounts for momentum loss, friction loss, and additional mixing losses.

(geometry changes) $$\Delta P = K(\rho V^2 / 2) \tag{5}$$

Applying this same concept to an exhaust system, total pressure drop across a catalytic converter can be correlated using a loss coefficient, which can be measured for either cold or hot flow. Fig. 7-33 shows the loss coefficient for a typical catalytic converter [7.34].

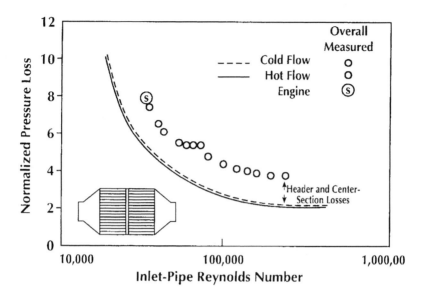

Fig. 7-33 Converter normalized pressure drop. (Source: Ref. [7.34].)

Loss coefficients based on Eq. 3 have also been established for correlating the pressure drop contributions of many types of piping sections [7.35], such as:

- Bends and mitres
- Dividing and joining joints
- Corners
- Abrupt entries and area changes

In all cases, the loss coefficient does not include the effects of gas pulsations.

Loss coefficients have been reported in the literature for many flow-passage geometries, and apply for Reynolds numbers larger than 2×10^6. Fig. 7-34 shows loss coefficients for an abrupt inlet expansion and an abrupt exit contraction, graphed as a function of the ratio of the converter frontal area to the frontal area of the inlet pipe. Velocity head for this correlation is based on the inlet pipe size. Loss coefficients for 90-degree bends with alternative corner treatments are presented in Fig. 7-35 [7.35].

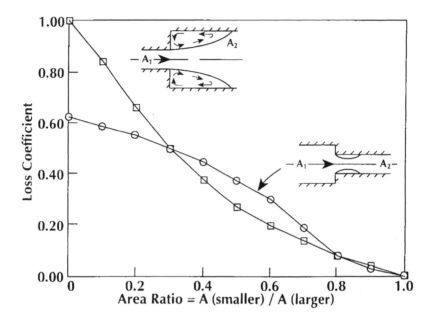

Fig. 7-34 Loss coefficients for abrupt contraction and expansion.

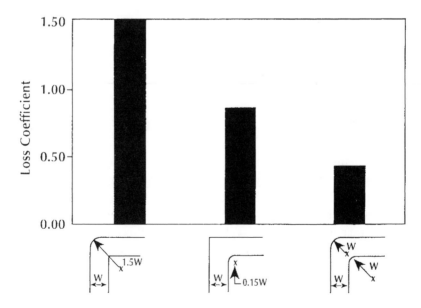

Fig. 7-35 Loss coefficients for a 90-degree bend.

Vehicle Applications

The contribution of individual pressure drops to the total pressure drop for the system is of considerable interest to the exhaust systems engineer. As described in the previous section titled "Exhaust System Pressures and Pressure Drop," these contributions can be summed to predict the total pressure drop and pressure distribution along a particular exhaust system.

Calculation of the overall pressure distribution for an exhaust system requires detailed knowledge of the geometries of all components, including exact the dimensions of pipe diameters, lengths, and bends. Loss coefficients must be known for converters, resonators, mufflers, and pipe joints and bends. In addition, a measured or calculated gas flow rate and temperature are required at the inlet to each component. The effects of augmentation must also be considered. There is evidence that loss coefficients are augmented for pipe joints and bends in an exhaust system because of flow dynamics, with the amount of augmentation depending on the location of the component in the exhaust system. An example of the pressure distribution in a complete exhaust system is presented in Fig. 7-36. This pressure distribution is typical of a mid-1980s vehicle powered by a four-cylinder engine and using a single underfloor converter. In this system, the converter, muffler, and piping each contribute approximately one-third of the total backpressure.

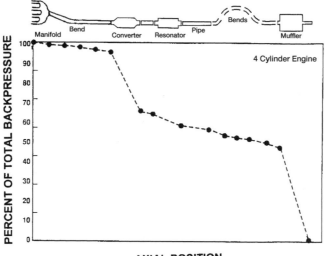

Fig. 7-36 Exhaust system pressure distribution.

Nomenclature

A	area
D	diameter
f	friction factor
K	pressure loss coefficient
L	flow length
p	pressure
Re	Reynolds number
V	velocity
w	exhaust gas mass flow rate
ρ	fluid density
D	difference

Subscripts

fr	frontal
lam	laminar
loss	loss
max	maximum
static	static
tot	total
u	uniform

References

7.1 U.S. Government, "Control of Air Pollution From Motor Vehicles and New Motor Vehicle Engines," Federal Register, Part II, Environmental Protection Agency, March 7, 1991.

7.2 Robertson, G.F., "A Study of Thermal Energy Conservation in Exhaust Pipes," SAE Paper No. 790307, Society of Automotive Engineers, Warrendale Pa., 1979.

7.3 Heck, R.H., Wei, J., and Katzer, J.R., "Mathematical Modeling of Automotive Catalysts," AIChE J., Vol. 22, p. 477, 1976.

7.4 Oh, S.H. and Cavendish, J.C., "Transients of Monolithic Catalytic Converters: Response to a Step Change in Feedstream Temperature as Related to Controlling Automobile Emissions," *Ind. Eng. Chem. Prod. Dev.*, Vol. 21, pp.29–37, 1982.

7.5 Chen, D.K.S., Bissett, E.J., Oh, S.H., and Van Ostrom, D.L., "A Three-Dimensional Model for the Analysis of Transient Thermal and Conversion Characteristics of Monolithic Catalytic Converters," SAE Paper No. 880282, Society of Automotive Engineers, Warrendale Pa., 1988.

7.6 Voltz, S.E., Morgan, C.R., Liederman, D., and Jacob, S.M., "Kinetic Study of Carbon Monoxide and Propylene Oxidation in Platinum Catalysts," *Ind. Eng. Chem. Prod. Res. Dev.*, No. 12, p. 294, 1973.

7.7 Chen, D.K.S., "A Numerical Model for Thermal Problems in Exhaust Systems," SAE Paper No. 931070, VTMS1, Society of Automotive Engineers, Warrendale Pa., 1993.

7.8 Seider, E.N. and Tate, C.E., "Heat Transfer and Pressure Drop of Liquids in Tubes," Ind. Eng. Chem., Vol. 28, p. 1429,1936.

7.9 Wendland, D.W., "Automobile Exhaust-System Steady-State Heat Transfer," SAE Paper No. 931085, VTMS1, Society of Automotive Engineers, Warrendale Pa., 1993.

7.10 De Sousa, E., "Dual Wall Energy Conserving Pipes for Internal Combustion Engines," SAE Paper No. 931095, VTMS1, Society of Automotive Engineers Warrendale Pa., 1993.

7.11 Hartsock, D.L., Stiles, E.D., Bable, W.C., and Kranig, J.V., "Analytical and Experimental Evaluation of a Thermally Insulated Automotive Exhaust System," SAE Paper No.940312, Society of Automotive Engineers, Warrendale Pa., 1994.

7.12 Chen, D.K.S. and Cole, C.E., "Numerical Simulation and Experimental Verification of Conversion and Thermal Responses for a Pt/Rh Metal Monolithic Converter," SAE Paper No. 890789, Society of Automotive Engineers, Warrendale Pa., 1988.

7.13 Moore, W.R. and Mondt, J.R., "Predicted Cold Start Emission Reductions Resulting from Exhaust Thermal Energy Conservation to Quicken Catalytic Converter Lightoff," SAE Paper No. 931087, VTMS1, Society of Automotive Engineers, Warrendale Pa., 1993.

7.14 Moore, W. and Myers J.P., "An Experimental and Analytical Heat Transfer Study of Takedown Pipes to Lower Cold Start HC Emissions," C496/085/95, VTMS2, Institution of Mechanical Engineers, London England, 1995.

7.15 Pattas, K.N., Stamatelos, A.M., Pistikopoulos, P.K., Koltsakis, G.C., Konstandinidis, P.A., Volpi, E., and Leverone, E., "Transient Modeling of 3-Way Catalytic Converters," SAE Paper No. 940934, Global Emission Technology and Analysis, SP-1043, Society of Automotive Engineers, Warrendale Pa., 1994

7.16 Mondt, J.R., "A Guard System to Limit Catalytic Converter Temperatures," SAE Paper No. 760320, Society of Automotive Engineers, Warrendale Pa., 1996.

7.17 Mondt, J.R., "Exhaust Aftertreatment Subsystems," The Challenge of Future Passenger Car Emissions Standards TOPTEC, Society of Automotive Engineers, Warrendale Pa., 1993.

7.18 Heck, R. and Farrauto, R., "Automotive Catalysts," *Automotive Engineering*, Vol 104, No. 2, 1996.

7.19 Ball, D.J., "Distribution of Warm-Up and Underfloor Catalyst Volumes," SAE Paper No. 922338, Society of Automotive Engineers, Warrendale Pa., 1992.

7.20 Terres, F. et al., "Electrically Heated Catalyst-Design and Operation Requirements," SAE Paper No. 961137, Society of Automotive Engineers, Warrendale Pa., 1996.

7.21 Engler, B.H. et al., "Reduction of Exhaust Gas Emissions by Using Hydrocarbon Adsorber Systems," SAE Paper No. 930738, Society of Automotive Engineers, Warrendale Pa., 1993.

7.22 Heimrich, M.J. et al., "Electrically-Heated Catalyst System Conversions of Two Current Technology Vehicles," SAE Paper No. 910612, Society of Automotive Engineers, Warrendale Pa., 1991.

7.23 Heimrich, M.J., et al., "Electrically-Heated Catalysts for Cold-Start Emission Control on Gasoline and Methanol-Fueled Vehicles," ICE-Vol. 15, Fuels, Controls and Aftertreatment for Low Emissions Engines, ASME 1991.

7.24 Socha, S.S., Jr. and Thompson, D.F., "Electrically Heated Extruded Metal Converters for Low Emission Vehicles," SAE Paper No. 920093, Society of Automotive Engines, Warrendale Pa., 1991.

7.25 Hellman, K.H. et al., "Evaluation of Different Resistively Heated Catalyst Technologies," SAE Paper No. 912382, Society of Automotive Engineers, Warrendale Pa., 1991.

7.26 Kubsh, J.E. and Lissiuk, P.W. "Vehicle Emission Performance with an Electrically Heated Converter System," SAE Paper No. 912385, Society of Automotive Engineers, Warrendale Pa., 1991.

7.27 Whittenberger, W.A. et al., "Experiences with 20 User Vehicles Equipped with Electrically Heated Catalyst Systems-Part I," SAE Paper No. 920722 (SP-910), Society of Automotive Engineers, Warrendale Pa., 1992.

7.28 Socha, L.S. et al., "Reduced Energy and Power Consumption for Electrically Heated Extruded Metal Converters," SAE Paper No. 930383 (SP-968), Society of Automotive Engineers, Warrendale Pa., 1993.

7.29 Baccarini, I.N., et al., "Data Analysis of Independently Run EHC Programs," SAE Paper No. 920850, Society of Automotive Engineers, Warrendale Pa., 1992.

7.30 Shimasaki, Y., et al, "Development of Extruded Electrically Heated Catalyst System for ULEV Standards," SAE Paper No. 971031, Society of Automotive Engineers, Warrendale Pa., 1997.

7.31 Hanel, F.J. et al., "Practical Experience with the EHC System in the BMW ALPINA B12," SAE Paper No. 970263, Society of Automotive Engines, Warrendale Pa., 1997.

7.32 Kaiser, F.W. et al., "Optimization of an Electrically-Heated Catalytic Converter System-Calculations and Application," SAE Paper No. 930384, Society of Automotive Engines, Warrendale Pa., 1993.

7.33 Shapiro, A.H., *The Dynamics and Thermodynamics of Compressible Fluid Flow*, Vol. 1, The Ronald Press, New York, N.Y., 1953.

7.34 Wendland, D.W. et al. "Sources of Monolith Catalytic Converter Pressure Loss," SAE Paper No. 912372, Society of Automotive Engineers, Warrendale Pa., 1991.

7.35 Miller, D.S., Internal Flow Systems, Vol. 5, BHRA Fluid Engineering Series, 1978.

Chapter 8

Alternative Fuels and Global Emissions

The U.S. automotive industry has always functioned in a complex economic, social, and political environment. During the past 30 years, the industry has become highly regulated, with standards imposed on noise and safety, fuel economy, and emissions. Much of this regulation is a direct result of scientific and public policy concerns about environmental air and water pollution, global warming (the greenhouse effect), and acid rain. Against this background, industry research has sought ways to lessen the impact of automotive waste products on the greater environment. Efforts in the late 1990s to control emissions from automobiles have focused primarily on developing alternative fuels; changing the composition of standard fuels; and developing effective catalysts, especially a lean-burn NO_X catalyst.

Alternative Fuels

With the passage of the Revised Clean Air Act of 1990, the auto industry realized that better fuels with more stringent controls on fuel content and quality would be needed [8.1]. The term "standard fuels" refers to gasoline and diesel fuels. "Alternative" fuels include those that are liquids at ambient pressure and temperature, and those that are gases at ambient temperature and pressure. Liquid fuels can be stored and distributed using essentially the same system already in place for gasoline and diesel fuels. Gaseous fuels, on the other hand, require large storage vessels if they are to be stored at ambient temperature and pressure; alternatively, they can be stored at high pressures, or as liquids at cryogenic temperatures. Compared to liquid fuels, all gaseous fuels are at a relative disadvantage in terms of storage and distribution.

Examples of liquid alternative fuels are M85 (a mixture of 85% methanol and 15% gasoline), methanol, ethanol, and reformulated gasoline. A popular reformulated gasoline contains the additive methyl tertiary butyl ether (MTBE), which is an oxygenated liquid that costs approximately 20% more to refine than gasoline. In mid-1999, MTBE was identified as a *possible* carcinogen. As a result, California banned MBTE from use in California beginning in the year 2003.

Oxygenated fuels are formed by replacing a hydrogen atom in the fuel molecule with either O or OH. A simple example is methanol, CH_3OH, where an OH molecule replaces and H molecule in methane, CH_4. This substitution results in oxygen being available within the fuel molecule to aid in the oxidizing reaction or combustion of the fuel. Replacing the hydrogen atom in a fuel with an oxidizing agent lowers the heating value per pound or volume. As a result, more fuel is required to provide an equivalent amount of energy. Another type of oxygenated fuel is dimethyl ether (DME). This fuel also has favorable characteristics for oxidizing HC and CO, but its cost is estimated to be at least 50% greater than that of diesel fuel [8.2].

The volumes of different fuels required to travel a fixed distance in an automobile, as compared to gasoline, are presented in Fig. 8-1. Fuels compared include diesel, DME, LNG (liquid natural gas), propane, ethanol, methanol, and CNG (compressed natural gas). These fuel volumes were estimated for a driving journey that includes some city driving as well some freeway driving. Data on fuel heating values and relative cycle efficiencies for this figure were obtained from available SAE references [8.2, 8.3] and are summarized in Table 8-1. Using an Otto cycle and gasoline fuel as the baseline, with a cycle efficiency of 1.0, relative cycle efficiencies were calculated for all alternative fuels, and were used to generate a "fictitious" fuel heating value to account for different thermodynamic cycle efficiencies. Based on a higher compression ratio and negligible intake throttling losses, the cycle efficiency for diesel and DME is 1.3, whereas it is 1.0 for gasoline, LNG, CNG, and propane. To account for higher octane fuel usage, cycle efficiencies for ethanol and methanol are slightly higher than those for gasoline.

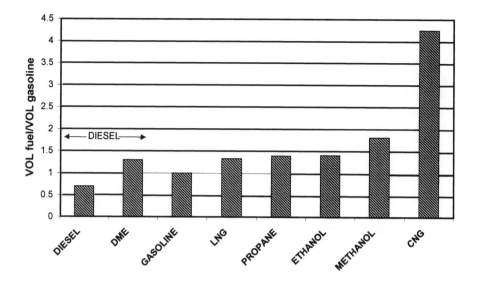

Fig.8-1 Relative fuel volume for equal distance traveled.

Table 8-1. Energy Comparisons for Alternative Fuels

	Heating Value (MJ/kg)	Density (kg/m³)	Heating Value (MJ/m³)	Relative Cycle Efficiency	"Fictitious" Heating Value (MJ/m³)
DIESEL	42.7	831	35484	1.3	46129
DME	28.8	667	19210	1.3	24972
GASOLINE	43.2	750	32400	1.0	32400
LNG	49.0	497	24353	1.0	24353
PROPANE	46.4	500	23175	1.0	23175
ETHANOL	26.4	789	20830	1.10	22913
METHANOL	20.0	795	18762	1.12	17808
CNG	49.0	Variable	37.3*	1.0	7612**

* at 1 atm pressure
** at 3000 psia pressure

Tradeoffs in terms of cold-start difficulty, availability, and cost of alternative fuels are compared in Fig. 8-2. In this figure, one arrow indicates a factor of 1.5 to 2.0 and two arrows indicate a factor of 2 to 10. It can be seen from this figure that reformulated gasoline lowers HC and CO emissions substantially for vehicles without catalytic converters. All alternative fuels listed likely will require a catalyst. For the hydrogen-powered vehicle, assuming it uses air as a source of oxygen, an NO_X catalyst will be needed.

Fuel	Exhaust Catalyst	Cold Start	Availability	Cost
Gasoline	TWC	—	—	—
Diesel	OX	—	—	—
DME	OX	—	↓↓	↑↑
Liquid				
M85	TWC	—	—	—
Methanol	TWC	↓	—	—
Ethanol	TWC	↓	↓	↑
Reformulated Gasoline (MTBE)	TWC	—	—	—
Gaseous				
Natural Gas (CNG)	TWC	—	—	—
Propane (LNG)	TWC	—	↓	↑
Hydrogen	NOx	—	↓	↑↑
Electric	—	↑	↓	↑↑

↑ 1.50 → 2.0
↑↑ 2.0 → 10.0

Fig. 8-2 Advantages, disadvantages, and emission controls for alternative fuels.

A comparison of emissions from these alternative fuels is presented in Fig. 8-3, in a format similar to that of Fig. 8-2. From this we can see that one disadvantage of oxygenated fuels is that they produce aldehydes, which are now a controlled exhaust emission; and a disadvantage of diesel engines is that they produce more particulates than spark-ignited engines. Overall, this comparison suggests that M85 may be a more promising fuel than gasoline, and that natural gas has much potential as an alternative fuel.

Fuel	HC	CO	NOx	Aldehydes	Particulates
Gasoline	—	—	—	—	—
Diesel Fuel	—	↓	—	—	↑
DME	—	↓	—	—	—
Liquid					
M85	—	—	—	↑	—
Methanol	↓	—	—	↑	—
Ethanol	↓	—	—	↑	—
Reformulated Gasoline (MTBE)	↓	↓ (open loop)	—	↑	—
Gaseous					
Natural Gas	↓	↓	—	—	↓
Propane	—	↓	—	—	↓
Hydrogen	↓↓	↓↓	—	↓↓	↓↓
Electric	↓↓	↓↓	↓↓	↓↓	↓↓

↑ 1.5 → 2.0
↑↑ 2.0 → 10.0

Fig. 8-3 Emissions comparisons for alternative fuels.

An example of using oxygenated fuels over a 20 year period to lower CO emissions was reported for Denver, Colorado [8.4]. In this study, the results of which are presented in Fig. 8-4, air quality measured in downtown Denver was used to indicate CO violation days. The average fleet emissions and CO violation days were correlated with specific years, showing significant lowering of both indicators of air quality.

Starting in 1972, major improvements in atmospheric levels of CO were recorded as emission controls were implemented for automobiles and other sources. Progress leveled off in 1980, however. When reformulated fuel, oxygenated gasoline, was required beginning in 1987, air quality again significantly improved. In 1987, Denver started requiring oxygenated fuel with oxygen content of 1.5% for all gasoline-fueled vehicles. Then, in 1988, the oxygen content was increased to 2.0%. From Fig. 8-4 we can see that CO violation days at the Denver downtown monitoring station were lowered from 120 in 1973 to 8 in 1988.

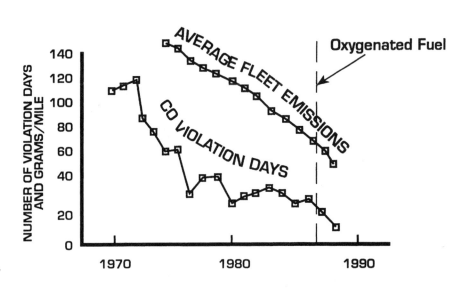

Fig. 8-4 CO violation days in downtown Denver vs. average tailpipe CO emissions. (Source: Ref. [8.4].)

The U.S. oil industry deserves recognition for years of dependable service supplying lubricant and gasoline fuels to the auto industry and the motoring public. However, as emissions regulations became more restrictive, it became obvious that the content and quality of gasoline fuels must be improved. Data generated by the Auto/Oil Air Quality Improvement Research Program (AQIRP), which is discussed below, helped provide impetus for improving gasoline fuels. The term used to describe these better fuels was "reformulated gasoline."

Reformulated gasoline encompasses a broad spectrum of fuel specifications, depending on the oil company supplying the gasoline. Suppliers of reformulated gasoline include Arco, Diamond Shamrock, Conoco, Phillips, Marathon, Mobil, and Shell. The properties of a reformulated gasoline should be seasonal, to accommodate temperature differences, and regional, to accommodate air quality demands. For example, Reid vapor pressure (RVP) should vary for satisfactory cold starts, and be low in summer compared to winter. However, in nonattainment regions such as the Los Angles Basin, to help reduce ozone formation the RVP should be no higher than 9.

Additional typical requirements for reformulated gasoline fuels include:

- Maximum allowed: aromatics 25 percent; benzene 0.8 percent; and olefins 5 percent.
- A sulfur content not exceeding 300 ppm, with an ultimate target of less than 50 ppm, for improved catalyst performance, especially durable performance.
- A 90 percent distillation temperature within the range of 140– 150°C (280–300°F).
- Oxygen content of at least 2.5 weight percent, especially to lower carbon monoxide emissions from older vehicles.

Auto/Oil Air Quality Improvement Research Program

Recognizing that fuels, engines, and emissions are inexorably entwined, in 1989 fourteen oil companies and the three domestic automobile manufacturers initiated the Auto/Oil Air Quality Improvement Research Program (AQIRP), the goal of which was to develop data on fuel/vehicle systems to help legislators, regulators, and companies meet the nation's clean air goals [8.5]. This study included three general subject areas: (1) extensive vehicle emission measurements, (2) air-quality modeling studies focused on ozone formation, and (3) economic analysis of several alternative fuel and vehicle systems.

The six-year program encompassed two phases and was completed after testing over 100 vehicles, using over 90 fuel compositions with more than 5000 emissions tests [8.6]. Vehicle tailpipe, evaporative, and running emissions were measured including NO_x , CO, and 151 volatile organic compounds (VOCs), which included all HC emissions. Fuel combinations covered a wide variation in aromatic and olefin content, oxygenate type and content, sulfur content, vapor pressure, and distillation temperature. Methanol and gasoline mixtures ranged from 0% methanol to 85% methanol (M85). Three vehicles were tested using four mixtures of natural gas having methane concentrations ranging from 86 to 97%.

For Phase I, the vehicle fleet consisted of fourteen older models (1983–85), equipped primarily with carburetors; and a fleet of 20 newer models (1989), equipped with prototpype advanced emission control technology. For Phase II, 25 additional vehicles were added: seven aged 1986–87 vehicles high emitters, six 1993 California models, six 1994 Federal models, and six 1990 prototypes with advanced emission controls. The 1989 fleet had accumulated a minimum of 10,000 miles of service, while the older fleet had accumulated 40,000 to 80,000 miles.

In May 1993, findings from program studies from 1989 through 1992 were published as a Phase I report. Then, in January 1997, a Phase II final report was published, covering all three general subject areas. Copies of these reports can be obtained from the Coordinating Research Council. Results from the Phase I and Phase II studies were also published in many SAE reports. An abstract of significant findings follows.

Fuel Compositional Changes

Effects on vehicle emissions of variations in fuel aromatics, oxygenates, T90, T50, sulfur, and RVP were measured.

Reduction of aromatics from 45% to 20%:

- lowered HC emissions by 6 to 14%, depending on the age of the vehicle;
- lowered CO emissions by as much as 13% for the current fleet;
- lowered NO_X emissions by 11% for older vehicles.

Adding approximately 2.7% by weight of oxygenates, such as MTBE or ethanol:

- lowered HC by 5–9%;
- lowered CO by 11–14%;
- had essentially no impact on NO_X

Reducing the amount of olefins in the fuel from 20% to 5%:

- increased HC by 6%;
- had no impact on CO, and lowered NO_X by 6%.

Lowering sulfur content in gasoline from 450 ppm to 50 ppm:

- lowered HC by 18%;
- lowered CO by 19%;
- lowered NO_X by 8%.

Lowering gasoline distillation points, either T90 from 325 to 280°F (163 to 138°C) or T50 from 215 to 185°F (102 to 85°C):

- lowered HC by 5–10%;
- had little impact on CO;
- increased NO_X by 5–11%

Reducing fuel volatility from 9 psi to 8 psi (61.2 to 54.4 kPa):

- decreased evaporative emissions;
- lowered HC by 4%;
- lowered CO by 9%;
- had little effect on NO_X.

Use of reformulated gasoline, CaRFG, a commercially available fuel that includes several of the above listed modifications:

- lowered HC by 10–27%;
- lowered CO by 21–28%;
- lowered NO_X by 7–16%;

Use of fuel with 85% methanol (M85):

- lowered HC by 31%;
- lowered CO by 13%;
- lowered NO_X by 6%.

Use of compressed natural gas (CNG):
- lowered HC by 80–90%;
- lowered CO by 40–80%;
- lowered NO_X by 10–80%.

Ozone Modeling

The Urban Airshed Model, originally developed by EPA and updated by AQIRP, was used to study smog components, especially the formation of ozone with vehicle and fuel changes. This model ultimately included exhaust, evaporative, and running emissions from vehicles, as well as refueling and fuel storage emissions from stationary sources. It incorporated recent information on atmospheric chemistry, effects of sunlight, and deposition of pollutants on surfaces. Four urban airsheds were modeled: New York City, Los Angeles, Dallas-Fort Worth, and Chicago. These cities were chosen because of their high ozone levels and variety of atmospheric conditions, but also because the information on these cities needed for the model was readily available.

Using the 1988 model as reference, the predictive model for 2010 was changed to incorporate federal low emission vehicles, improved reformulated gasoline fuel, enhanced vehicle inspection/maintenance programs and estimates of controls on stationary sources. Using estimates for the future, the model predicts that the contribution to peak ozone from light-duty vehicles in the four cities modeled will decrease from 28–37% in 1980–85 to only 5–9% by the years 2005–2010. This predicted decrease in ozone is consistent with the decrease in HC, CO, and NO_X emissions contributed by light-duty vehicles as older high emitters are retired. (Contributions to HC, CO, and NO_X emissions from light-duty vehicles are reported more completely in Chapter 9.)

The AQIRP study represents a major cooperative research effort by the automobile and fuel industries toward advancing understanding of the contributions of vehicles and fuels to atmospheric pollution. Data generated has been used to establish specifications for improved fuels, now referred to as "reformulated fuels," to aid in controlling automotive emissions. The program has also advanced significantly the scientific knowledge of atmospheric models, which predict future decreased exhaust and evaporative emissions from vehicles. The overall result of AQIRP has been a reduction in both ozone formation and toxic emissions in urban areas.

Global Warming (Greenhouse Effect)

Global warming, also called the "greenhouse effect," refers to the impact on the earth's temperature from accumulated gases in the troposphere. At this time there are many uncertainties about global warming and the greenhouse effect.

The greenhouse effect is this: The sun supplies energy to the earth in the form of radiated ultraviolet (UV) and visible light. Upon reaching the earth, this energy is converted to infrared (IR) radiation, and is redirected toward outer space. If there were no gases in the troposphere, this redirected energy would be reflected back into space. However, when gases are present in the troposphere, some of this energy is reflected back to the earth's surface, increasing the surface temperature (Fig. 8-5).

The delicate balance between energy radiated into space and that reflected to the earth's surface results in an average temperature of the earth surface of 15°C (59°F) instead of a very cold –20°C (–4°F). However, if gases in the troposphere cause too much energy to be reflected back to the earth, the

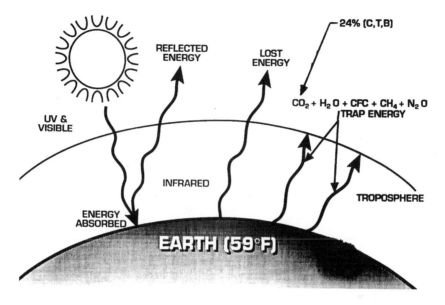

Fig. 8-5 Global energy balance greenhouse effect.

surface temperature will increase to uninhabitable levels unless unwanted gases are prevented from adding to the layer already in place. Of all the gases in the troposphere, CO_2 and chlorinated fluorocarbons (CFCs) such as Freon, are primary gases from automobiles that contribute to the greenhouse effect; in 1993, U.S. cars and light trucks contributed less than 2.0% of the CO_2 [8.7].

Freon gas reflects back to earth a significant fraction of thermal energy from the stratosphere. Prior to 1975, Freon was the cooling fluid of choice for automotive air conditioning systems. Replacing Freon, CFC-12 (chlorofluo-rocarbon-12) three times during the life of an auto air conditioning system has approximately the same impact on the environment as the CO_2 and NO_X produced by the same vehicle driven 100,000 miles. Beginning in 1975, CFC-12 refrigerant in all new cars was replaced by HFC-134a (hydrofluorocarbon, 1-1-1-2 tetrafluoroethane). In addition, when aged cars are retired, Freon is recovered from auto air conditioning systems and reprocessed. By using HFC-134a, which does not contain chlorine compounds, in place of Freon, the contribution of automotive air conditioning systems to the greenhouse effect is being minimized.

In its 1992 annual report, the Energy Information Agency of the U.S. Department of Energy identified CO_2, CH_4, and NO_2 as the three major "greenhouse" gases. Compared to CO_2, methane is estimated to be 20 times more effective at reflecting thermal energy and nitrogen dioxide more than 290 times more effective. Data for atmospheric CO_2 measured at the Mauna Loa Observatory in Hawaii showed an increase in atmospheric CO_2 from 315 to 340 ppm in the time period 1958 to 1980 (Fig. 8-6). This increase resulted from an imbalance in the earth's ecosystem caused by the inability of plants, trees, and bushes to convert the CO_2 back to elemental oxygen and carbon.

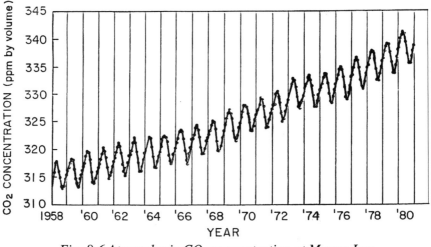

Fig. 8-6 Atmospheric CO_2 concentration at Mauna Loa Observatory, Hawaii.

In the five years prior to 1992, emissions of CO_2 from natural sources totaled about 160,000 million metric tons per year, with human activities contributing an additional 8000 million tons. It is estimated that natural mechanisms can remove about 165,000 metric tons per year, thus producing a net increase in CO_2 of 3000 million metric tons. Similarly, the annual increase in methane is estimated to be about 30 million metric tons per year, and that of NO_X about 4 million metric tons per year.

In 1986, approximately 24% of the CO_2 in the troposphere was supplied by combustion in the U.S. of all fossil-fuels [8.3]. Alternative fuels have been proposed to help alleviate the "greenhouse effect," by reducing the amount of CO_2 produced. The relative amounts of CO_2 produced by the combustion of alternative liquid and gaseous fuels are shown in Fig. 8-7 [8.4].

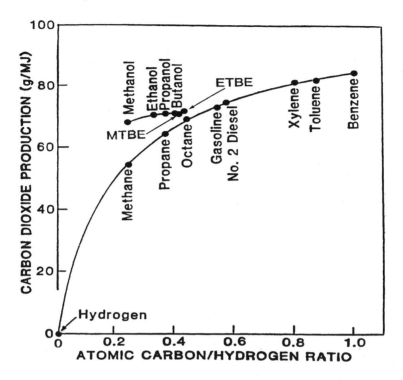

Fig. 8-7 Carbon dioxide production by alternative fuels.
(Source: Ref. [8.3].)

Although this comparison is valid only for vehicle powerplants with the same thermal efficiency, the findings are useful. On a grams per equivalent energy (g/Mj) basis, production of carbon dioxide is altered very little through the use of proposed alternative liquid fuels. However, complete combustion of natural gas, which is mostly methane, offers a significant gain of approximately 21%. Of course, the ultimate fuel for reducing the amount of CO_2 produced is hydrogen, which contains zero carbon. Although not a fuel, electric power is also a candidate, assuming the electricity is produced by nuclear- power stations.

Improvement in vehicle fuel consumption leads to a proportionate lowering of CO_2. For example, using a diesel engine with 20% improvement in fuel consumption will lower CO_2 production by 20%. Better fuel economy, resulting in less CO_2, can be obtained by using a lean-burn gasoline engine or a diesel engine. However, the diesel engine currently faces emission control challenges, especially with regard to particulate and NO_X emissions.

205

The European community has indicated much concern about the greenhouse effect and lowering CO_2 produced by vehicles. Since over 40% of the vehicles in Europe are powered by diesel engines, the desire to control NO_X produced by diesel engines is a high priority. Throughout the industrialized world, the desire to improve efficiency and control NO_X from lean-burn and diesel engines has prompted aggressive programs to modify the combustion space geometry, improve injection systems, and develop improved fuels. In addition, investigations are underway to develop lean-burn NO_X catalysts.

Lean Burn NO$_X$ Catalyst Consortium

In 1993, the U.S. federal government and the U.S. automotive industry joined in an historic partnership to establish global leadership in the development and production of affordable, fuel-efficient, low-emission automobiles. This alliance, the Partnership for a New Generation of Vehicles (PNGV), drew on a wide range of resources: Included were 8 federal agencies, national laboratories, universities, and automotive suppliers, and the U.S. Council for Automotive Research (USCAR), a pre-competitive, cooperative research effort between General Motors, Ford, and Chrysler. To facilitate joint programs among industries and government agencies, the U.S. Congress established a Cooperative Research and Development Activity (CRADA). Under the terms of a CRADA, a cooperative research effort between U.S. industry and the national laboratories was funded at one or more of the national laboratories by an agency of the U.S. Government. In return for this federal funding, the companies agreed to support the effort with matching "work-in-kind," or an equivalent amount of materials, man-hours, and testing. The primary objective of the PNGV program was to make American industry more globally competitive with foreign manufacturing companies.

One of the programs organized as part of USCAR, as a critical enabling technology for the commercialization of lean-burn engines, was an effort called "Reduction of Nitrogen Oxide Emissions for Lean Burn Engine Technology." A cooperative research and development agreement was negotiated in 1993 between the three U.S. auto companies and Lawrence Livermore Laboratories, Los Alamos National Laboratory, Oak Ridge National Laboratory, and Sandia National Laboratory to simultaneously study potential lean-burn catalysts [8.8]. A plethora of very advanced technologies, including aerogels, hydrous metal oxides, perovskites, and zeolites were investigated as part of this R & D effort toward lean-burn catalysts. During the three-year

$ 6.5 million program, these technologies were evaluated, through analytical and experimental methods, for possible use in automotive catalysts. In addition, a modern engine dynamometer facility to measure catalyst performance was built at the Oak Ridge, Y-12 Laboratory.

At the conclusion of the three-year program, two of the potential technologies that were tested at Oak Ridge were actually incorporated into catalytic converters. The performance of these experimental prototypes essentially matched the performance of experimental prototypes available from commercial catalyst companies. After the initial three-year effort came to a close, a scaled-back version of the program was continued with the aim of eventually incorporating some of this advanced technology into production vehicle systems.

To reduce NO_X requires the presence of a reducing agent, such as a hydrocarbon, in the highly oxidizing exhaust gas stream. Propylene has been found to be one of the most active reducing agents, although other hydrocarbons will work as well. By injecting propylene into the exhaust stream, maximum conversion efficiencies of 30 to 40% have been obtained for the two most promising lean-burn NO_X catalysts, copper zeolite (Cu/ZSM5) and platinum catalyst on alumina. These results are shown in Figs. 8-8 and 8-9. Unfortunately, these efficiencies are possible for only a "narrow" temperature window, depending on the catalyst. For a Pt catalyst on alumina, the window is between 280 and 310°C, and for Cu/ZSM5, it is between 370 and 420°C. A viable NO_X catalyst should have a much higher efficiency, at least 80%, and a broader temperature window. Catalyst companies, both within the U.S. and overseas, have continued to aggressively pursue research and development on lean-burn catalysts to increase efficiency and broaden the temperature window.

An alternative to a lean-burn NO_X catalyst is a "storage catalyst"; a type of catalyst that adsorbs NO_X during lean operation and then cycles to a rich condition for a very short time, allowing a three-way catalyst to reduce the stored NO_X. Results from experimental efforts indicate that maintaining the catalyst at a rich condition for a very short duration is sufficient to chemically reduce the NO_X. However, a special catalyst formulation is needed that does not chemically react with the adsorbed NO_X. Barium in the washcoat, which retains the NO_X as a nitrate, has proven a good adsorbing medium. However, sulfur in fuels combines with the barium to form a sulfate, making

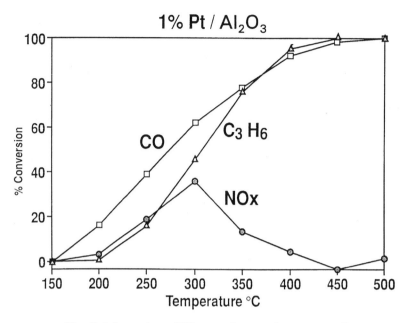

Fig. 8-8 Lean-burn NO_X catalyst performance—Pt.

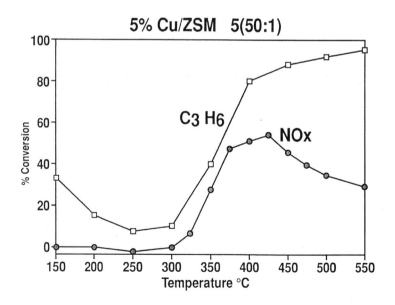

Fig. 8-9 Lean-burn NO_X catalyst performance—Cu/ZSM5.

it unsuitable for NO_X adsorption. Another method under study is that of injecting additional fuel into the exhaust to provide the reducing hydrocarbons needed, instead of cycling to a rich A/F.

Acid Rain

Acid rain is the term used to describe the undesirable deposition of minute particles of acidic material that originate in the atmosphere from sulfur dioxide (SO_2) and oxides of nitrogen (NO_X). When SO_2 and NO_X react with gases in the atmosphere, they produce sulfuric acid and nitric acid, making the resulting rainfall more acidic (Fig. 8-10). Acid rain can have adverse effects on sensitive lakes, streams, soils, and forests. It is included in this discussion because it is a form of environmental pollution. However, it has been found that cars and trucks contribute only an estimated 2% of the sulfur dioxide and 30% of the NO_X precursors to the atmosphere in the 31 states east of or bordering the Mississippi River (Fig. 8-11). By far, most of the sulfur dioxide and NO_X precursors in the atmosphere are contributed by industrial manufacturing plants, power plants, and chemical plants.

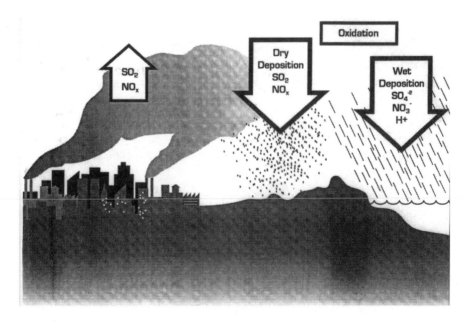

Fig. 8-10 Acid rain.

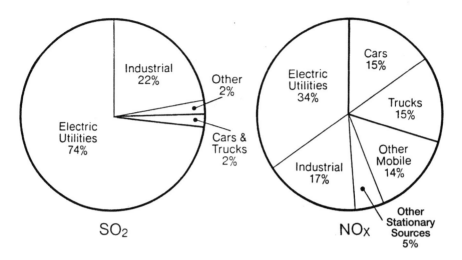

Fig. 8-11 Sources of sulfur dioxide and oxides of nitrogen emissions

Forty to sixty percent of the constituents are deposited into the atmosphere as dry particles. The remaining amount is oxidized and is deposited as "acid" rain. Acid rain has a pH between 4 and 5, compared to normal rain, which has a pH between 5 and 5.5 (neutral pH is 7) [8.9]. Because weather patterns in the Northern Hemisphere predominately move from west to east, the northeastern United States and the regions of eastern Canada are the prime recipients of acid rain originating in the industrial states located to their west.

References

8.1   Colucci, J.M., "What Can the Oil Industry do in the 1990's—An Auto Man's Perspective," GMR Publication 7010, Presented to API Forum, 1990.

8.2   Verbeek, R. and Van der Weide, J., "Global Assessment of Dimethyl-Ether: Comparison with Other Fuels," SAE Paper No. 971607, Society of Automotive Engineers, Warrendale Pa., 1997.

8.3   Amann, C.A., "The Passenger Car and the Greenhouse Effect," SAE Fuels and Lubricants Meeting, Tulsa, Okla., 1990.

8.4   Gallagher, J., Livo, K.B., Hollman, T., and Miron, W.L., "The Colorado Oxygenated Fuels Program," SAE Paper No. 900063, Society of Automotive Engineers, Warrendale Pa., 1990.

8.5   AQIRP, "Auto/Oil Air Quality Improvement Research Phase I Final Report," May 1993.

8.6   AQIRP, "Auto/Oil Air Quality Improvement Research Phase I Final Report," January 1997.

8.7   GM Corporation, "Focused on Improving Competetiveness," Public Interest Report, 1993

8.8   United States Council for Automotive Research, "PNGV Technical Accomplishments, 1996," USCAR, U.S. Department of Energy, 1996.

8.9   GM Corporation, "General Motors Public Interest Report," 1983.

Chapter 9

Future Perspective

Since the mid-1960s, the automotive industry has done a remarkable job of engineering systems to control emissions from automobiles. A measure of the industry's success is the fact that, as the twenty-first century begins, tailpipe emissions of HC, CO, and NO_X will have been reduced by 99%, 96%, and 95%, respectively, compared to 1965 levels. And all of this will have been accomplished with vehicles that have much improved fuel economy; ride and handle better; are safer; and, on the relative basis of wage earnings to inflation, are of reasonable cost.

Fuel Quality

To support the auto industry in its quest to control automotive emissions, the oil companies have made significant contributions by improving fuel quality; however, in the future, they will be required to do more. Oil companies will be under pressure to lower the sulfur content of gasoline and diesel fuel to 50 ppm. For gasoline, the target for aromatics is 35% by volume. Improved gasoline quality and content can be expected, with improvements including the use of reformulated, oxygenated gasolines. Some regions in the U.S. are already making good use of these gasolines, at a slightly higher cost per gallon. The use of oxygenated fuels will likely increase, although for the very low emissions vehicles now being produced, data suggest that oxygenated fuels do not contribute much to an additional decrease in emissions.

Alternative Fuels

In 1995, approximately 150 million vehicles were licensed in the United States, consuming approximately 200 million gallons of gasoline per day. Meeting this demand for fuel represents an awesome challenge to any potential alter-

Retail Cost of Emission Controls

Incorporating new technology into motor vehicles to lower emissions means extra costs to the consumer. An estimate of the cumulative additional cost for the installation of emission controls on a typical passenger car is presented in Fig. S9-1. These data, based on the AAMA Statistical Summary for 1997, are for the time period 1968–1997. Note that the cost increase parallels very closely the severity of emissions regulations, presented in Fig. 6-1 on page 81.

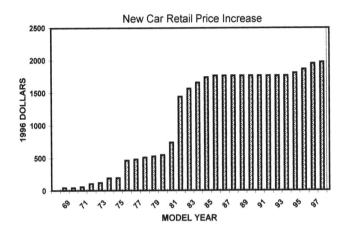

Fig. S9-1 Cumulative cost of emission controls. (Data source: Ref. [S9.1].)

The addition of oxidizing converter systems in 1975 added approximately $220 to the cost of a typical passenger vehicle. And, when three-way catalytic converter systems were implemented in 1980–83, the additional cost increment amounted to approximately $1200. With allowances for inflation, this figure has been carried forward on each annual model change. Most recently, to meet the lowered emissions standards required by the Revised Clean Air Act of 1990, a modest cost of approximately $200 was added to the cumulative cost of equipping a vehicle with a state-of-the-art emission control. All told, the cumulative additional cost of equipping a vehicle with state-of-the-art emission control systems that met the mandated standards was approximately $2000 for a 1997 vehicle.

Reference

S9.1   American Automobile Manufacturers Association, "Motor Vehicle Facts and Figures, 1997," AAMA, Washington, D.C., 1997.

native fuel. Obviously, the logical alternative fuel would be a liquid that could be accommodated in the storage and distribution system that already exists for gasoline and diesel fuel. However, since no alternative fuel can be expected to suddenly supplant gasoline as the fuel of choice, some duplication of equipment likely will be necessary.

The Clean Air Act of 1990 requires fleets of vehicles to convert to alternative fuels. Bus fleets, for example, will probably be the first group to make the full conversion to fuels such as natural gas. Meanwhile, much technological progress is being made in modifying diesel engines for use with alternative fuels such as natural gas.

The use of alternative fuels, other than gasoline and diesel, will steadily increase. In the late 1990s, only a small number of vehicles—mostly in fleet service—use alternative fuels, predominantly liquid natural gas and mixtures of methanol and gasoline, such as M85. More and more, liquid natural gas is becoming the fuel of choice because the exhaust products from its combustion have low reactivity in ozone-producing reactions.

Emission Standards

There is no evidence that the relentless pursuit of improved air quality will subside. Air quality has become a concern worldwide, and the control of pollutants from automobiles and all other sources has become a priority environmental objective. During the 1990s, most of the world's industrialized nations began to establish emission controls for all sources, including automobiles. The1998 Honda is one example of a system reflecting the current state of the art; it meets ULEV emissions for a small vehicle powered by an L4 engine [9.1]. This system integrates four technologies:

- Levels of HC are lowered during the cold start by control
  ling the A/F ratio at a lean value. Variable valve timing and
  lift are important during the cold start.
- The exhaust manifold and exhaust pipe are air-gap pipes
  designed with thin inner liners to conserve exhaust sensible
  heat, to quickly heat the catalyst.
- An improved catalyst formulation provides high-conversion
  performance.
- The fuel control system maintains precise control of the A/F
  ratio.

Emissions data for the Honda small car were reported as 0.03, 0.35, and 0.12 g/mi for NMOG, CO, and NO$_X$, respectively, for a system aged to 100,000 miles. Honda claims that this vehicle will provide cleaner air at the tailpipe than the ambient air mix ingested into the engine at a typical intersection in metropolitan Tokyo.

With the auto companies demonstrating vehicles that are approaching ULEV, California in 1998 proposed additional emission regulations, identified as LEVII, ULEVII, and SULEV (super-ultra-low-emission vehicles), to become effective shortly after the year 2000. Several of the proposed emission standards are summarized in Table 9-1 [9.2, 9.3]. In addition, in 1999 California proposed that SUVs (sport utility vehicles) be required to meet the same tailpipe emissions standards as passenger cars. This was prompted in part by the increase in the number of these vehicles being purchased by the motoring public.

Table 9-1. Proposed Exhaust Emission Standards

Category	Durability, miles	NMOG (g/mi)	CO (g/mi)	NO$_x$ (g/mi)	PM (g/mi)
LEVII	50,000	0.075	3.40.	050.	0.08
ULEVII	100,000	0.04	1.7	0.05	0.04
SULEV	120,000	0.010	1.0	0.02	0.01

By the year 2000, vehicle emission standards will be in effect in at least 25 nations worldwide. The European Community and Japan already have standards for automobile emissions which closely parallel those currently in force in the United States. While other countries continue to struggle with technologies, manufacturing facilities, and world trade regulations, an estimated 95% of all new vehicles in the next several years will come equipped with some type of emission controls, with many systems using a catalyst. For the foreseeable future, continued lowering of emissions standards will impact all fossil-fuel-burning sources of power, placing increased demands on the automotive industry.

Emissions Control Progress

Requirements mandated by the Revised Clean Air Act of 1990 continue to be implemented. Changes to the U.S. driving schedule have been prompted by new emissions regulations for vehicles traveling at high speeds with increased loads and with air conditioning. To control emissions from vehicles operating at loads higher than required for the US FTP driving test, an additional driving schedule, the US06 schedule, has been developed (Fig. 9-1). This schedule had been added to the US FTP and has its own set of emissions standards: 10g/mi for both US Tier I and CARB; and HC + NO_X of 0.8 for the U.S. and 0.15 for CARB.

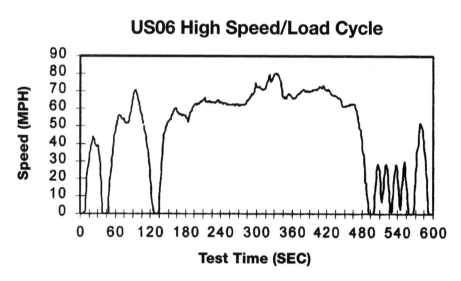

Fig. 9-1 US06 high-speed driving schedule.

Air conditioning must also be included in testing using a "modified" FTP, the AC02 test cycle. Modifications include a more strenuous first five cycles, followed by the standard remaining 13 cycles, as shown in Fig. 9-2. The air conditioning system must be operating for this AC02 driving cycle, as well as for the US06 test. In both cases, radiant heating is used to simulate a sun load with an ambient temperature of 95° F and humidity of 40%. Emissions measured with the air conditioning operating are combined with weighting factors to determine if they meet the legislated standards promulgated by the Revised Clean Air Act of 1990. Both the high-speed and air conditioning regulations were targeted to phase in beginning in the year 2000 as the Supplemental Federal Test Procedure (SFTP).

AC02 A/C Test Cycle
Speed vs Time

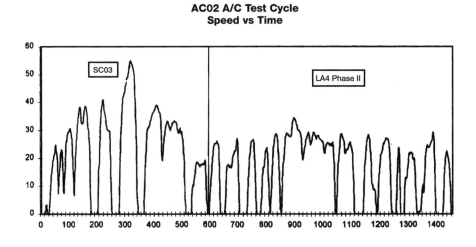

Fig. 9-2 Air conditioning driving schedule.

Air Quality Data

Data gathered by the American Association of Automobile Manufacturers (AAMA) documents improvement in air quality in the U.S. relative to that measured beginning in 1940 [9.4]. In 1995, according to AAMA, on-road vehicles contributed 27.2%, 62.3%, and 31.9% of the total atmospheric VOCs, CO, and NO_X, respectively [9.4]. The vehicles in the study included gasoline and diesel cars, trucks, and motorcycles.

The total emissions of VOCs in the U.S. and the contribution of gasoline-powered vehicles are indicated in Fig. 9-3. The data show that reductions in automotive emissions have contributed most of the decrease in VOCs from the peak year of 1970, and that the total now is approaching that measured in 1940. Similarly, Fig. 9-4 shows the total CO emissions in the U.S. and the

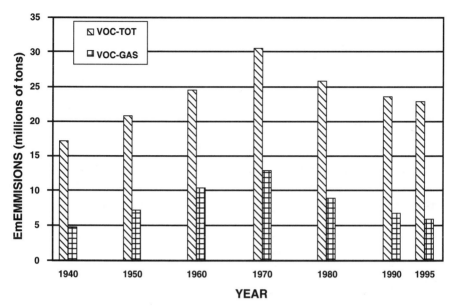

Fig. 9-3 Total VOCs and VOCs from gasoline-powered vehicles. (Data source: Refs. [9.3] and [9.4].)

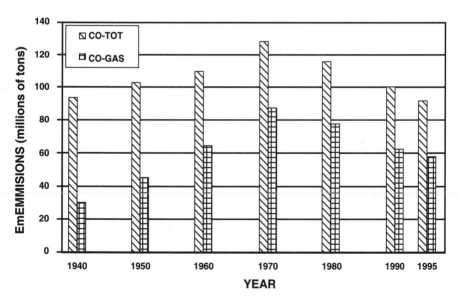

Fig. 9-4 Total CO and CO from gasoline-powered vehicles. (Data source: Refs. [9.4] and [9.5].)

contribution of gasoline-powered vehicles to that total over the same time span. CO also peaked in 1970 and has been declining through 1995. The total CO now is also approaching 1940 levels.

Unfortunately, changes in NO_X emissions are not as encouraging; total NO_X reached its highest level in 1980, dropped off somewhat, but then began to increase again, as shown in Fig. 9-5. Contribution to total NO_X from gasoline-powered vehicles peaked in 1980, and then declined somewhat in subsequent years, but not as much as levels of HC and CO. This disparity provides the justification for regulators to target sources of NO_X, which include all fossil fuel burning engines (automobile, aircraft, off-road, farm tractor; small engines, and motorcycles) as well as fuel consuming industries.

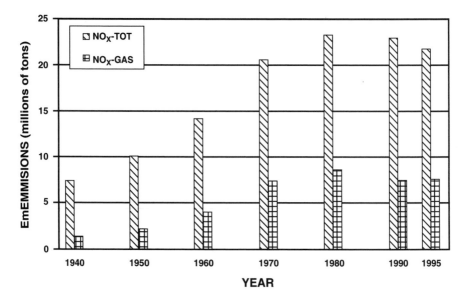

Fig. 9-5 Total NO_X and NO_X from gasoline-powered vehicle. (Data source: Refs. [9.4] and [9.5].)

Total emissions from vehicles in the United States have decreased significantly as emission control systems have become more efficient. Total VOC emissions from vehicles peaked in approximately 1970, as shown in Fig. 9-6.

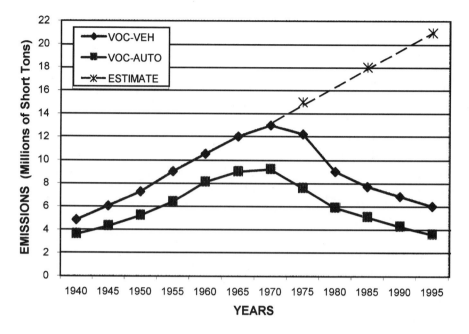

Fig. 9-6 VOC Emissions from vehicles. (Data source: Refs. [9.4] and [9.5].)

VOC emissions have been reduced to the extent that they now approach the values attributed to vehicles in 1940—and this decrease has come about even though the vehicle fleet has increased from 25 million in 1947 to 80 million in 1970 and 125 million in 1996. Between 1940 and 1996, total miles driven by all vehicles has increased from 300 billion to 2470 billion. Far and away, the largest portion of the decrease in emissions is attributed to passenger cars. As shown in Fig. 9-7, total CO emissions from U.S. vehicles show a trend similar to that for VOCs. However, total CO emissions does not approach the 1940 values as closely as the value for total VOCs does.

Total NO_X emissions from all vehicles peaked in approximately 1980 and declined slightly until 1990, as shown in Fig. 9-8. The decrease in NO_X emissions can be attributed primarily to a decreased output from passenger cars; however, the decrease is much less than that for either VOC or CO emissions. Since NO_X is a primary contributor to smog and ozone production, these data raised concern among environmentalists.

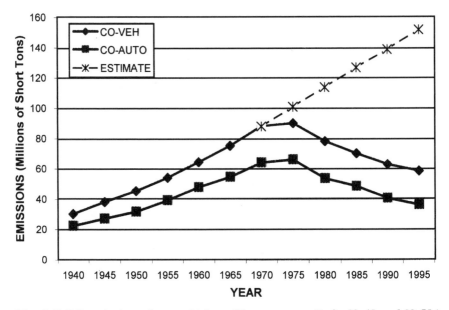

Fig. 9-7 CO emissions from vehicles. (Data source: Refs. [9.4] and [9.5].)

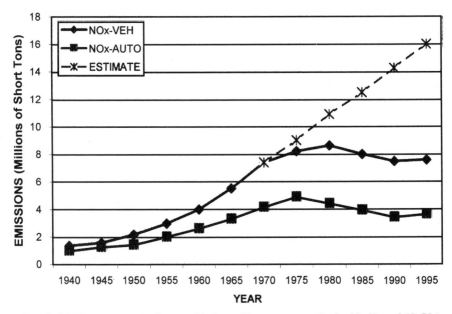

Fig. 9-8 NO$_X$ emissions from vehicles. (Data source: Refs. [9.4] and [9.5].)

Future Projections

As the worldwide community moves into the twenty-first century, the automobile is becoming more and more important as an element of our "improved" standard of living. As more countries build and operate automobiles, control of emissions and fuel efficiency will be mandatory. As additional improvements are made to existing spark-ignited engines, the demand for improved fuel economy will dictate diesel or diesel-like engines in order to reduce emissions of CO_2 into the atmosphere and mitigate global warming and the "greenhouse effect."

New technologies on the horizon, some of which were operational in limited production in the 1990s, include inlet supercharging, variable valve timing, variable stroke, and direct fuel injection [9.6]. These techniques can lead to significant improvements in fuel economy—as much as 20%, depending on the engine operating A/F ratio. Transmission and driveline efficiencies and an infinitely variable transmission (IVT) can permit engine operation closer to peak efficiency at all times, further improving fuel economy.

By combining supercharging with cam phasing, engines can be operated more efficiently, with potential improvement in fuel economy of 15%. High-pressure, intermittent fuel injection has shown the potential to improve fuel economy and lower emissions. Continued efforts to reduce friction include the use of roller cam followers. And more exact machining of cylinder bores ensures that when the engine is hot, a "round" shape is retained to mach the piston's "round" shape.

For spark-ignition engines operating at stoichiometric A/F, improved three-way catalysts will be required to lower emissions to ULEV levels; and for lean-burn or diesel engines operating at lean A/F, improved lean-burn NO_X catalyst systems will be required.

A diesel cycle engine offers the potential, for a 35–40% improvement in fuel economy as compared with a spark-ignited engine for the same size vehicle, and an increase in fuel economy translates directly to lower CO_2 and less greenhouse gas. For these reasons, the diesel engine is a desirable powerplant for vehicles. However, control of particulates and NO_X emissions from diesel engines poses a formidable challenge to the diesel engine designer, the fuel industry, and emission control engineers. Better fuel mixing with higher-pressure sprays, multiple valves, and variable displacement are all potential improvements to engine designs. Improved fuel specifications for cetane number and sulfur content, for example, are being implemented to improve diesel fuels. In addition, diesel fuel additives to lower particulates are being investigated.

Various catalyst approaches to reduce NO_X, even in highly oxidizing gas mixtures, are under development. An approach with much promise is controlled cycling of the A/F from very lean to a stoichiometric value, enabling a three-way catalyst combined with an adsorber to reduce NO_X. To be successful, this approach requires a catalyst and washcoat mix that efficiently adsorbs NO_X during lean A/F operations and also functions as an efficient oxidizing catalyst for HC and CO control. The catalytic reduction of NO_X is performed during very short excursions to a rich A/F. To maintain an acceptable operation of the engine, the amount of NO_X storage must be monitored and fuel control adjusted accordingly to precisely cycle the A/F between the normal lean condition to a rich A/F.

Electric vehicles will offer some contribution to lowering vehicle emissions, but their impact is not expected to be substantial until the ever-elusive efficient, small, lightweight, long-life battery is developed. Although many battery prototypes have been studied, the nickel-metal-hydride seems the best choice to date. General Motors EV1 vehicle, leased for customer evaluation in California, Arizona, and New Mexico in 1996, originally used lead-acid batteries for energy storage; and advanced version of the vehicle is being upgraded with nickel-metal hydride batteries. In 1998, Toyota offered consumers in Japan an electric vehicle, the Prius, also powered by nickel-metal hydride batteries; this to be available for sale in the United States in 2000. Further, Suzuki is experimenting with an electrically powered commercial vehicle, CT-1, which again uses nickel-metal hydride batteries [9.7].

In 1993, the U.S. Government established the Partnership for a New Generation of Vehicles (PNGV), discussed previously in Chapter 8. The major objective of PNGV has been to demonstrate an automobile that will comfortably transport four adults, achieve an efficiency of 80 miles per gallon (33.8 km/L), and be affordable [9.8]. The U.S. auto companies involved in the PNGV program continue to pursue their in-house development efforts, which include such promising future vehicle powerplants as hybrids and fuel cells.

Because of concerns about global warming, worldwide interest in lowering CO_2 emissions has placed additional emphasis on development efforts involving hybrids of fuel-burning engines and electric drives. A variety of hybrid combinations are possible, including diesel, gasoline, gas turbine, and Stirling engines, in series or in parallel with electric drive motors. A number of concept vehicles have been built to demonstrate the various technologies. GM has reported developing a gas turbine engine combined with electric drive motors, as well as a parallel hybrid powered by an electric motor and a direct injection turbo engine. Honda has announced a series hybrid using a direct-injection, in-line three-cylinder engine and a CVT connected to an electric motor. Toyota has reported a conventional spark-ignited internal-combustion engine in a parallel hybrid powerplant.

Vehicles powered by fuel cells are also being developed. Hydrogen can be used as fuel when it is combined with oxygen in an electrolytic process that produces electricity and does not require combustion. A reformer uses a catalyst to strip hydrogen from either methanol or gasoline, whichever is the primary fuel provided to the vehicle. The hydrogen is then ducted to the fuel cell, where it reacts with oxygen from ambient to ultimately produce electricity for an electric motor to drive the vehicle. The only "emission" from this process is water. However, a byproduct of hydrogen stripping is the production in the reformer of CO and CO_2 gases, which can poison the catalyst and destroy its ability to strip the hydrogen.

Fuel efficiencies several times greater than those available for a conventional internal combustion engine have been demonstrated in a variety of fuel cell systems. Ford, Daimler-Benz, and Ballard Battery have joined forces with the aim of producing a fuel-cell-powered car by 2004. Chrysler advertised a "proof-of-concept" fuel cell prototype of 1999. (The merger of Daimler Benz with Chrysler will probably affect the status of these programs.) General Motors has declared that it will have a "production ready" fuel-cell car by 2004. Estimates of the cost of a fuel-cell "engine" vary from two to three times the cost of a typical gasoline internal combustion engine. However, the fuel cell as a hybrid combination with batteries and electric drives may very well be the automotive powerplant for the next century.

Although it is impossible to predict the future, it appears that there remains a significant future for the internal combustion engine. Until at least 2010 or 2020, the "improved" internal combustion engine, either gasoline or diesel, will likely continue to be the favorite engine for vehicles before being chal-lenged seriously by an exotic new powerplant. The hybrid appears to be the closest challenger on the horizon, with the fuel cell not too far behind.

The twenty-first century promises to be an exciting, challenging, and occasionally frustrating time for the automotive engineer, Meeting society's demands for improved air quality with low pollution, while continuing to provide customer with comfortable, efficient, driveable, and affordable vehicles will remain a formidable task. Despite the current regulatory environment and the presence of worldwide competition, the future remains bright for creative, dedicated, industrious engineers who are not intimidated by technical challenges. As the history and development of emission control has shown, there is great professional satisfaction in working toward and finally achieving a designated goal, no matter how difficult the journey. And there is special satisfaction in knowing that the goal is providing improved air quality for mankind.

References

9.1 Kishi, N., Kikuchi, S., Seki, Y., Kato, A., and Fujimori, K., "Development of the High Performance L4 Engine LEV System," SAE Paper No. 980415, Society of Automotive Engineers, Warrendale, Pa., 1998.

9.2 Walsh, M.P., "Global Trends in Diesel Emissions Control—A 1998 Update," SAE Paper No. 980186, Society of Automotive Engineers, Warrendale, Pa., 1986.

9.3 CARB, "Proposed Amendments to California Exhaust, Evaporative and Refueling Emissions Standards and Test Procedures for Passenger Cars, Light-Duty Trucks and Medium-Duty Vehicles—LEVII," Preliminary Draft Status Report, June 1998.

9.4 AAMA, "Motor Vehicle Facts and Figures, 1995," Washington, D.C., 1995.

9.5 AAMA, "Motor Vehicle Facts and Figures, 1996," Washington, D.C., 1996.

9.6 Walzer, P., "Future Trends in Automotive Engine Technology," Automotive Technology International, 1997.

9.7 Yamaguchi, J., "Concept Vehicles and Technology at the Tokyo Motor Show," Automotive Engineering International, January 1998.

9.8 Miller, V. and Beeber, R., "PNGV Technical Accomplishments," U.S. Department of Commerce, 1996.

Appendix A

Timeline for Control of Automotive Emissions in the United States

Year	Title	Comment
1946	South Coast Air Quality Management District	Formed by California legislation to control pollution in Los Angeles basin.
1947	Los Angeles County Air Pollution District	Formed by California legislation.
1960	California Motor Vehicle Control Board	Established by California legislation.
1961	PCV California	Provided by auto companies, prior to legislation.
1963	PCV USA	Provided by auto companies, prior to legislation.
1963	Clean Air Act for USA	Congress passed first legislation controlling environmental pollution.
1965	Amended Clean Air Act	Auto emissions added to controls.
1966	First year for emission controls in California	Response to Clean Air Act.
1967	Amended Clean Air Act	Air Quality Act by U.S. Congress.
1967	Inter-Industry Emission Control Program (IIEC1)	Formed to develop an emission-free automotive powerplant.

Year	Title	Comment
1968	California Air Resources Board (CARB)	Established by California legislation.
1968	Federal Emission for light duty vehicles for 1970	Established by HEW for tailpipe and evaporative emissions.
1968	First year for emission controls in USA	Response to Clean Air Act.
1968	CARB empowered to establish emission standards for diesel-powered vehicles	Established by California legislation.
1968	7-Mode Driving Test cycle	Established by CARB, emission test.
1969	End-of-line audit procedure	Established by CARB.
1969	Smog Case antitrust suit against GM, Ford, Chrysler, and American Motors	United States Department of Justice under the Sherman Act.
1970	Environmental Protection Agency (EPA) established	Established by U.S. Congress, William Ruckelshaus appointed administrator.
1970	Clean Air Act amended, "Muskie Bill."	U.S. Congress legislation setting future emissions targets, and 50,000 mile durability, target 1975.
1970	National Ambient Air Quality Standards (NAAQS)	EPA established first air quality standards.
1972	Lead removal from gasoline established by EPA	0.05 g/gal, upper limit by 1975, no leaded fuel after January 1, 1997.
1972	Original Federal Test Procedure (FTP), cold start only	Established by EPA in 1970, Driving Test Procedure, replaced the 7-Mode Test.

Year	Title	Comment
1973	Exhaust gas recirculation (EGR) introduced	Required to meet EPA NO_X standards.
1973	Regulations for diesel-powered light duty vehicles	Established by CARB, dynamometer test.
1974	IIEC2 formed	Inter-Industry Emission Control Program extended.
1975	Oxidizing catalysts introduced	Required to meet EPA standards.
1975	75FTP, Revised Federal Test Procedure	EPA revised test procedure to include cold start and hot start.
1975	First Corporate Average Fuel Economy (CAFE) standards	Established by U.S. Congress, the Energy and Conservation Act of 1975.
1977	IIEC2 terminated	
1981	Three-way catalysts and electronic controls introduced	Required to meet EPA standards.
1981	1963 Clean Air Act expired	Requirements extended until 1990.
1986	Particulate standard introduced	Established by EPA.
1987	Chrysler purchased American Motors	
1989	Auto/Oil Air Quality Improvement Program (AQIRP)	Established by 14 oil companies and 3 U.S. automakers.
1990	Revised Clean Air Act	U.S. Congress, sweeping of 1990 changes to requirements for emission controls.
1992	Ozone Transport Commission (OTC)	Established by 12 northeastern states plus the District of Columbia.

Year	Title	Comment
1993	Formaldehyde standard introduced	Established by EPA.
1997	Auto/Oil Air Quality Final Report	Report covering all three original phases.
1994	U.S. vehicles began phasing in TLEV, LEV, ULEV, and ZLEV	Response to 1990 Revised Clean Air Act.
1998	Daimler Benz and Chrysler merge	
1999	National Low Emissions (NLEV) Vehicles	Established by EPA, phasing in of LEV standards in 9 OTC states.

Appendix B

People and Places

Many, many individuals have contributed to this book. The first group consists of the many outstanding individuals whom I have had the opportunity to work with and for during my automotive career. Obviously, my contacts have primarily been with individuals within General Motors, where I worked for 43 years. The second group consists of individuals from the Ford and Chrysler organizations some of whom I have not had the opportunity to meet. To honor their contributions to the development of automotive emission controls, I have asked a colleague from each organization to comment on significant events in the development of automotive emission controls in their "shop" and the individuals involved. The third group consists of members of the supplier community, especially catalyst companies. I would like to acknowledge them for their contributions to advancing the state of the art in emission control systems.

General Motors

Beginning in the late 1960s, General Motors mounted a monumental effort to develop systems and components to control emissions from automobiles. This effort was a logical extension of GM's ongoing research and development activities, and ultimately involved all of GM's operations.

In 1969, at GM Research, a team of chemists, physical chemists, and physicists was assembled in the Physical Chemistry Department to focus on the possible use of catalytic converters to control automotive emissions. Preliminary studies were encouraging, but available catalysts were sparse and existing converters were much too large. Thus, a team consisting of members of the Research Laboratories, Engineering Staff, Buick Division, and AC Spark Plug Division was assembled to pursue advanced systems and technologies to lower tailpipe emissions.

General Motor's management designated the AC Spark Plug Division to serve as the focal point for all catalysts supplied to General Motors. This eliminated the confusion and duplication of effort that would have delayed the development of emission control technologies; the various catalyst suppliers no longer had to coordinate programs with many different GM car divisions. AC Spark Plug Division also developed screening tests to expedite sorting and rating performance of alternative catalyst materials from suppliers.

A staff of catalyst experts was assembled at the Research Laboratories. John Larson, head of the Physical Chemistry Department, served as leader of this staff, which included Richard Klimish, Kathy Taylor, Louis Hegedus, Jerry Summers, James Schlatter, and others.

The Fuels and Lubricants Department was led by Charles Tuesday, and staff included William Agnew, Joseph Colucci, Norman Brinkman, James Spearot, Fred Bowditch, Jack Benson, and many others. The smog chamber was developed in the Fuels and Lubricants Department through the efforts of Joe Wentworth, Charles Begeman, Joe Collucci, Chuck Tuesday, and John Kaplan.

The Emissions Research and Engine Research Departments focused their efforts on studying engine controls and alternative engine designs. William Agnew, Charles Amann, Nick Gallopolous, and James Mattavi, with the support of many colleagues, contributed toward developing the technology necessary to control emissions from automobiles.

The GM Proving Grounds was an invaluable resource, testing hundreds of vehicles thousands of miles, to provide emission control data. R. Johnson, M. Homfeld, and W. Kolbe were early contributors to vehicle test procedures and test schedules. Al Robinson consistently advocated common sense approaches to vehicle emission testing and correlation of data. Harold Haskew contributed his knowledge of fuels and evaporative emissions, as well as his skills in negotiating with government agencies for logical solutions to emission controls.

During the 1960s, technical experts and administrators from the various GM staffs, primarily Research, had to spend a great deal of their time testifying in Washington, D.C., responding to requests from the U.S. Government for information on progress in research and development toward meeting emission standards. This heightened federal interest in industry procedures prompted GM in 1971 to organize a new staff to deal with environmental regulations.

The Environmental Activities staff was responsible for interactions with all government agencies issuing any type of environmental regulations that concerned GM; this included not only motor vehicles regulations, but also plant emissions regulations. One of the primary responsibilities of the Environmental Activities staff was to improve GM's technical image with the public and the government, particularly with regard to environmental issues. The new vice president lured to head up this staff was Dr. Ernest Starkman, a member of the faculty of the University of California, whose specialty was combustion processes and associated chemistry.

Ford Motor Company

(Courtesy of Bob McCabe, Principal Staff Engineer, Chemical Engineering Department, Ford Research Laboratory)

The Ford Motor Company, under the leadership of Henry Ford II, also relied on its Research Laboratories to guide development of emission control systems. Starting out with fewer resources than General Motors, Ford entered into cooperative efforts with other industry supporters and suppliers, including oil companies and catalyst companies. An outgrowth of these joint efforts was the IIEC Program.

Inter-Industry Emission Control (IIEC) Program

Robert Campau, Executive Engineer, emerged as the spokesman for the IIEC effort, which involved eleven separate companies. Started in 1967, the three-year program, estimated to spend $7.0 million, ultimately spent $21.0 million over a six-year period.

Catalytic Converter Development

A team of chemists, physical chemists, engineers, and supporting staff was organized at the Ford Scientific Research Laboratories (SRL) to investigate catalysts and catalytic converter systems. J.T. Kummer, M. Shelef, and Klaus Otto contributed to the technology, and H.S.Gandhi established himself as a worldwide authority on catalyst systems for automobiles. Much of the work at the Ford Scientific Research Laboratories was carried out under the direction of Serge Gratch.

In addition to the work at SRL, notable contributions were made by Eugene Weaver and Jim Gagliardi in Powertrain Operations. These two individuals, together with Joe Kummer, are generally credited with Ford's decision to proceed with the development and application of monolithic catalytic converters, which are now the industry standard. The issue of monoliths vs. pellets for substrates, and which was the better choice, will probably never be satisfactorily answered; each had both advantages and disadvantages.

Chrysler Corporation

(Courtesy of Michael Brady, Engineering Specialist, Supervisor, Catalyst Development, Advanced Engine System Development, Chrysler Corporation.)

During the 1970s, Chrysler Corporation endured some very difficult times. Many industry observers did not believe the company would survive. The story of Lee Iaccoca's leadership and the company's return to financial health is a legend in the auto industry. During the trying times, Chrysler was in no position to mount a large effort to study emission controls. Nevertheless, a number of dedicated individuals, including Maxwell Teague, Floyd Allen, Richard Goodwillie, and Bernard Robertson, managed to ensure that technology to control emissions was incorporated into vehicles. Much of the technology was a product of joint projects between Chrysler and its supplier network, which historically has provided Chrysler with more of its emission control technology than that provided to General Motors and Ford.

In 1970, various parts of Chrysler were assigned the joint task of meeting the dictates of the Clean Air Act of 1970. In the research office, the Chemical Research Department was directed to develop a catalyst, and Power Plant Research, headed by James Franceschina, was assigned the task of integrating catalytic converters into vehicles. In 1970, the Chief Research Scientist for Basic Sciences was Clayton Lewis. Lewis was succeeded by D. Maxwell Teague, who headed the catalyst research effort through most of the formative years.

The head of Catalyst Development was Leo B. Clougherty, and the early group leaders were Jack Engel and Philip J. Willson, followed in the mid-'70s by Edward J. Lesniak, Philip J. Willson, and Michael J. Brady. Willson

234

did much of the early substrate development and selection for Chrysler, while Engel and Brady were responsible for formulation development; Lesniak was responsible for process development.

Chrysler worked with Universal Oil Products (UOP) to jointly develop catalyst formulation and design and build the manufacturing plant in Tulsa, Oklahoma. During the time period 1972–1973, the 1975 production oxidation catalyst systems were developed. Richard E. Goodwillie was the manager of the department, which included both the certification group and the early engine and emission system development group. At that time, Richard Geiss and Roger Orteca were the senior engineers for the development groups that were most actively involved with catalysts. They both reported to William Hoffmeier, who, in turn, reported to Goodwillie. Douglas Teague and Richard Geiss were involved with supplier interactions and selections.

By 1975, Floyd Allen and Bernard Robertson had assumed supervision of the development of the three-way catalyst system with feedback control, which was introduced in production in 1980 in California. Advanced Engine Systems Development was formed, with Floyd Allen, Richard O. Schaum, and Richard Geiss leading the efforts of Clinton L. Syverson, Dewane D. Cogswell, Galen Kerns, and David C. VanRaaphorst. The Engine Performance Development Department, headed by Howard Padgham, was responsible for carburetor development and dynamometer testing, of catalyst performance.

In 1980, Gordon Rinschler became manager of Advanced Engine Systems Development, with both groups reporting to James Franceschina. In recent years, suppliers have become more involved than in the 1970s; however, emission control system have always been calibrated within Chrysler.

Appendix C

Acronyms

A/F	air to fuel ratio
AAMA	American Association of Automobile Manufacturers
ADP	alternative durability procedure
AIR	air injection reactor
AMA	Automobile Manufacturers Association
AQIRP	Air Quality Improvement Research Program
BSFC	brake specific fuel consumption
$BSNO_X$	brake specific oxides of nitrogen
CAFE	Corporate Average Fuel Economy
CARB	California Air Resources Board
CFC	chlorofluorocarbon
CLA	chemiluminescence analyzer
CNG	compressed natural gas
CRADA	Cooperative Research and Development Activity
CRC	Coordinating Research Council
CVS	constant volume sampler
CVT	continuously variable transmission
DF	deterioration factor
DME	dimethyl ether
DOT	Department of Transportation
DS-VDV	distributor-spark vacuum delay valve
DS-VMV	distributor-spark vacuum modulator valve
ECE	Economic Commission for Europe
ECM	electronic control module
ECS	evaporation control system
EEC	European Economic Community
EFE	early fuel evaporation

EGR	exhaust gas recirculation
EGR-TVS	exhaust gas recirculation-thermal vacuum switch
EHC	electrically heated converter
EOS	exhaust oxygen sensor
EPA	Environmental Protection Agency
EPCA	Energy Policy and Conservation Act
EUDC	Extra Urban Driving Procedure
EVAP	evaporation control
FID	flame ionization detector
FTP	Federal Test Procedure
GVW	gross vehicle weight
H/C	hydrogen to carbon ratio
H_2S	hydrogen sulfide
HC	hydrocarbons
HEW	Health Education and Welfare
IIEC	Inter-Industry Emission Control
LEV	low emission vehicle
LPG	liquid petroleum gas
LTR	lean thermal reactor
MBT	minimum spark advance for best torque
MIL	malfunction indicator light
MPI	multiport fuel injection
MTBE	methyl tertiary butyl ether
MVPCB	California Motor Vehicle Control Board
NAAQS	National Ambient Air Quality Standards
NDIR	non-dispersive infrared
NLEV	National Low Emission Vehicle
NMOG	non-methane organic gases
NO_X	oxides of nitrogen ($NO + NO_2$)
OBD	on-board diagnostic
OEM	original equipment manufacturer
OTC	Ozone Transport Commission
PCV	positive crankcase ventilation
PM	particulate matter
PNGV	Program for New Generation Vehicle
PULSAIR	pulse air injection reactor

RAF	reactivity adjustment factor
RTR	rich thermal reactor
RVP	Reid vapor pressure
SEA	Selective Enforcement Audit
SFTP	Supplemental Federal Test Procedure
SIP	State Implementation Plan
SO_2	sulfur dioxide
SOF	soluble organic fraction
SPI	sequential port fuel injection
SULEV	super ultra-low emission vehicle
TBI	throttle body injection
TCS	transmission control spark
THERMAC	thermal air cleaner
TLEV	transition low emission vehicle
TVS	thermovacuum switch
TWC	three-way catalyst
ULEV	ultra-low emission vehicle
USCAR	United States Council for Automotive Research
ZEV	zero emission vehicle

Index

Abbreviations are used after the page number to indicate figure (f),
end note (n), or tabular material (t).

About the Author

J. Robert Mondt is a retired GM engineer with 43 years in the industry and 28 years experience designing, building, and testing exhaust systems for spark-ignition engines. He is recognized as an international authority on systems and hardware for controlling automobile exhaust emissions.

From 1988–1991, Mondt worked in the AC Rochester Division developing the metal-monolith catalytic converter. In 1990, he was awarded the Charles McCuen Special Achievement Award at the GM Research Laboratories for inventing and developing for production the "herringbone" corrugated geometry for heat exchangers and catalytic converters. From 1991–1996, he served as Staff Engineer, supervisor for the Exhaust Systems Technology and Consortia Management Group, Exhaust Subsystems Engineering, for Delphi Energy and Engine Management. He also was Exhaust Subsystems Coordinator for USCAR LEP Cold Start Emissions Team and Lean NO_X CRADA.

From 1969–1976, Mondt was a member of the original "GM Catalyst Team," joint with AC Spark Plug Division, and charged with developing a catalytic converter system to control emissions from spark-ignited piston engines. Prior to that he worked at GM Research, focusing on the application of thermodynamics, heat transfer, and fluid mechanics to research on advanced automotive powerplants including gas turbine, free pistons, hybrids, steam, and advanced gasoline and diesel. He was responsible for much basic research on gas turbine regenerator heat exchangers, and for the design and development of the steam generator for the GM Steam Car, circa 1970.

Mondt holds 15 patents, and has authored 20 publications, mostly through ASME, SAE, and AIChE, on heat transfer, heat exchangers, and emission control. He has made presentations throughout the world, including in

Brazil, Italy, England, Germany, Luxembourg, and India, and has lectured at various universities, including Stanford, Michigan State, Colorado State, University of Wisconsin, and Rensselaer Polytechnic Institute.

Mondt is a member of ASME, SAE, the Combustion Institute, the Michigan Catalysis Society, and Sigma Xi. He has served on the ASME Gas Turbine Heat Transfer Committee, and is currently a member of the ASME ad hoc Committee on Heat Transfer Education. In 1967, ASME recognized the "Mondt number" as a new dimensionless parameter for correlating conduction loss in high-performance heat exchangers, especially gas turbine regenerators and Stirling engine regenerators.

Mondt is a member of the SAE Vehicular Heat Transfer Activity, and initiated the Vehicle Thermal Management Systems Conferences, a joint effort between SAE and IMechE, and chaired the technical program committee for the first meeting, VTMS1, in 1993. Mondt has also served as chairman or co-chair for VTMS2, VTMS3, and VTMS4. He also serves as chairman of the SAE Transactions Selection Committee for all SAE papers on exhaust emissions.

Mondt was born and raised on a ranch in Colorado. He graduated from the University of Denver in 1953 with a BSME and immediately began employment at General Motors Research Laboratories. From 1954–1956, he was on military leave from GM, serving as 1st Lt., Ordnance Corps. Mondt earned a GM Fellowship to Stanford University, and graduated with an MSME in 1957. Following this he returned to GM Research.